Innovative Research in the Food Packaging to Improve Food Quality and Shelf Life

Innovative Research in the Food Packaging to Improve Food Quality and Shelf Life

Editors

Valeria Rizzo
Giuseppe Muratore

MDPI • Basel • Beijing • Wuhan • Barcelona • Belgrade • Manchester • Tokyo • Cluj • Tianjin

Editors

Valeria Rizzo
Department of Agriculture, Food and Environment (Di3A)
University of Catania
Catania
Italy

Giuseppe Muratore
Department of Agriculture, Food and Environment (Di3A)
University of Catania
Catania
Italy

Editorial Office

MDPI
St. Alban-Anlage 66
4052 Basel, Switzerland

This is a reprint of articles from the Special Issue published online in the open access journal *Foods* (ISSN 2304-8158) (available at: www.mdpi.com/journal/foods/special_issues/innovative_packaging).

For citation purposes, cite each article independently as indicated on the article page online and as indicated below:

LastName, A.A.; LastName, B.B.; LastName, C.C. Article Title. *Journal Name* **Year**, *Volume Number*, Page Range.

ISBN 978-3-0365-2637-9 (Hbk)
ISBN 978-3-0365-2636-2 (PDF)

Contents

About the Editors vii

Preface to "Innovative Research in the Food Packaging to Improve Food Quality and Shelf Life" ix

Valeria Rizzo, Sara Lombardo, Gaetano Pandino, Riccardo N. Barbagallo, Agata Mazzaglia, Cristina Restuccia, Giovanni Mauromicale and Giuseppe Muratore
Active Packaging-Releasing System with *Foeniculum vulgare* Essential Oil for the Quality Preservation of Ready-to-Cook (RTC) Globe Artichoke Slices
Reprinted from: *Foods* **2021**, *10*, 517, doi:10.3390/foods10030517 1

Jawad Sarfraz, Anlaug Ådland Hansen, John-Erik Haugen, Trung-Anh Le, Jorunn Nilsen, Josefine Skaret, Tan Phat Huynh and Marit Kvalvåg Pettersen
Biodegradable Active Packaging as an Alternative to Conventional Packaging: A Case Study with Chicken Fillets
Reprinted from: *Foods* **2021**, *10*, 1126, doi:10.3390/foods10051126 21

Noor L. Yusof, Noor-Azira Abdul Mutalib, U. K. Nazatul, A. H. Nadrah, Nurain Aziman, Hassan Fouad, Mohammad Jawaid, Asgar Ali, Lau Kia Kian and Mohini Sain
Efficacy of Biopolymer/Starch Based Antimicrobial Packaging for Chicken Breast Fillets
Reprinted from: *Foods* **2021**, *10*, 2379, doi:10.3390/foods10102379 41

Marit Kvalvåg Pettersen, Julie Nilsen-Nygaard, Anlaug Ådland Hansen, Mats Carlehög and Kristian Hovde Liland
Effect of Liquid Absorbent Pads and Packaging Parameters on Drip Loss and Quality of Chicken Breast Fillets
Reprinted from: *Foods* **2021**, *10*, 1340, doi:10.3390/foods10061340 61

Teresa Fernández-Menéndez, David García-López, Antonio Argüelles, Ana Fernández and Jaime Viña
Application of PET/Sepiolite Nanocomposite Trays to Improve Food Quality
Reprinted from: *Foods* **2021**, *10*, 1188, doi:10.3390/foods10061188 79

Flavia Dilucia, Valentina Lacivita, Matteo Alessandro Del Nobile and Amalia Conte
Improving the Storability of Cod Fish-Burgers According to the Zero-Waste Approach
Reprinted from: *Foods* **2021**, *10*, 1972, doi:10.3390/foods10091972 93

Valentina Lacivita, Anna Lucia Incoronato, Amalia Conte and Matteo Alessandro Del Nobile
Pomegranate Peel Powder as a Food Preservative in Fruit Salad: A Sustainable Approach
Reprinted from: *Foods* **2021**, *10*, 1359, doi:10.3390/foods10061359 107

Stefania Toscano, Valeria Rizzo, Fabio Licciardello, Daniela Romano and Giuseppe Muratore
Packaging Solutions to Extend the Shelf Life of Green Asparagus (*Asparagus officinalis* L.) 'Vegalim'
Reprinted from: *Foods* **2021**, *10*, 478, doi:10.3390/foods10020478 119

Marwa R. Ali, Aditya Parmar, Gniewko Niedbała, Tomasz Wojciechowski, Ahmed Abou El-Yazied, Hany G. Abd El-Gawad, Nihal E. Nahhas, Mohamed F. M. Ibrahim and Mohamed M. El-Mogy
Improved Shelf-Life and Consumer Acceptance of Fresh-Cut and Fried Potato Strips by an Edible Coating of Garden Cress Seed Mucilage
Reprinted from: *Foods* **2021**, *10*, 1536, doi:10.3390/foods10071536 135

Elena Arena, Valeria Rizzo, Fabio Licciardello, Biagio Fallico and Giuseppe Muratore
Effects of Light Exposure, Bottle Colour and Storage Temperature on the Quality of *Malvasia delle Lipari* Sweet Wine
Reprinted from: *Foods* **2021**, *10*, 1881, doi:10.3390/foods10081881 . . . . **149**

About the Editors

Valeria Rizzo

Valeria Rizzo received her Ph.D. in Food Science and Technology (2008), and she is a Food Technologist.

Her professional skills and scientific research activity cover different fields of the food technologies, from chemical and qualitative analysis of nutritional compounds to food processing, from packaging systems to the overall quality and safety of food products. She worked in different research centers at University, CRA (Council for Agricultural Research) and CNR (National Research Council of Italy).

In the academic years 2017/19, she carried out teaching activity in the master degree course in Food Science and Technology in the University of Catania.

Her research activity has produced more than 50 publications on qualified national and international journals and on congress proceedings. Her interest in food research is a natural attitude that she expressed in scientific reading and writing activities; she is a referee for various international scientific journals.

Giuseppe Muratore

Giuseppe Muratore is associate professor in 07/F1 Food Science and Technology, in the AGR/15 sector–Food Science and Technologies, at the Department Di3A–Section of Food Technologies, University of Catania since 2015. Since the academic year 1997/98, he has carried out an intense teaching activity; as teacher of "Food Packaging" in the master degree course in Food Science and Technology (LM70) and "Process and Sensory Evaluation of Food" in the degree course in Food Science and Technology (L26). From 2016 to 2020, he was the President of the Food Science and Technology course (L26).

His professional skills and scientific research activities cover different areas of food technology with particular reference to food processing, packaging systems and their influence on the quality of food products, he is the author of over 200 scientific publications.

He is a member of Collegio docenti of International postgraduates PhD course in "Agricultural, Food and Environmental Science", and of several associations.

Preface to "Innovative Research in the Food Packaging to Improve Food Quality and Shelf Life"

Food packaging and shelf life have been the subject of remarkable research in recent years. They are so important because only by understanding a good storage system is it possible to avoid any food waste. Moreover, the best packaging has to prolong the food quality while also reducing the packaging volume or better, become itself biodegradable, and guarantee the nutritional characteristics of food products.

In particular, the increasing interest in reducing packaging wastes is becoming a rising problem, just considering that food packaging alone contributes to a huge portion of total packaging wastes in the world. On the other side, consumers judge the food quality based on appearance and freshness, but also using their awareness of the environmental implications of packaging. Nowadays, many technologies can be applied to improve food quality and shelf life, such the application of edible films or coatings, from biodegradable materials or biopolymers, trying to reduce the package barrier requirements, incorporating natural bioactive compounds and lengthening shelf life making packaging easily compostable.

The present Special Issue is aimed at gathering cross-disciplinary approaches to increase food quality and shelf life working on innovative processing or materials for food packaging in order to provide very new solutions to producers.

The editors particularly appreciate the efforts of all the authors who participated in this Special Issue.

In particular, the editors are very proud to have been able to collect fascinating papers from three international research groups, two of those reporting on experimental trials on chicken fillets. Two other papers are focused on innovative trays and pads for active packaging, and finally two other papers linked with the interesting as innovative "zero-waste" and "sustainable" approach to improve shelf life.

Studies on active packaging, shelf life and the effect of light exposure on a sweet wine were the Editors' research group's contributions.

The Editors wish you an enjoyable lecture.

Valeria Rizzo, Giuseppe Muratore
Editors

Article

Active Packaging-Releasing System with *Foeniculum vulgare* Essential Oil for the Quality Preservation of Ready-to-Cook (RTC) Globe Artichoke Slices

Valeria Rizzo, Sara Lombardo, Gaetano Pandino, Riccardo N. Barbagallo, Agata Mazzaglia, Cristina Restuccia *, Giovanni Mauromicale and Giuseppe Muratore

Di3A, Dipartimento di Agricoltura, Alimentazione e Ambiente, University of Catania, via S. Sofia 100, 95123 Catania, Italy; vrizzo@unict.it (V.R.); sara.lombardo@unict.it (S.L.); g.pandino@unict.it (G.P.); rbarbaga@unict.it (R.N.B.); agata.mazzaglia@unict.it (A.M.); g.mauromicale@unict.it (G.M.); g.muratore@unict.it (G.M.)

* Correspondence: crestu@unict.it

Abstract: Two globe artichoke genotypes, "Spinoso sardo" and "Opera F1", have been processed as ready-to-cook (RTC) slices and refrigerated at 4 °C for 12 days (i) to evaluate the suitability to be processed as RTC slices; (ii) to evaluate the effect of a *Foeniculum vulgare* essential oil (EO) emitter, within an active package system, to delay quality decay, thus extending shelf life; (iii) to estimate the impact of EO emitter on the sensory profile of the RTC slices after cooking. Results revealed that both globe artichoke genotypes possess a good attitude to be processed as RTC product. "Opera F1" showed the best performances for color parameters, texture and chemical indexes, while "Spinoso sardo" showed lower mass loss (ML) over the storage time. The addition of EO emitter slowed down the consumption of O_2, better preserved texture when compared to the control and more effectively control polyphenol oxidase (PPO) activity and antioxidants' retention during the cold storage. Microbial counts in control globe artichoke RTC slices were significantly higher than those packed with EO emitter, confirming the inhibiting role played by EO of *F. vulgare*. In addition, the EO emitter did not influence negatively the sensory profile of RTC globe artichoke slices after microwave cooking.

Keywords: essential oil emitter; globe artichoke genotype; quality parameters; microbial growth; antioxidants' retention

Citation: Rizzo, V.; Lombardo, S.; Pandino, G.; Barbagallo, R.N.; Mazzaglia, A.; Restuccia, C.; Mauromicale, G.; Muratore, G. Active Packaging-Releasing System with *Foeniculum vulgare* Essential Oil for the Quality Preservation of Ready-to-Cook (RTC) Globe Artichoke Slices. *Foods* **2021**, *10*, 517. https://doi.org/10.3390/foods10030517

Academic Editor: Joaquín Gómez Estaca

Received: 13 January 2021
Accepted: 20 February 2021
Published: 2 March 2021

1. Introduction

Food market demand is increasingly focused on food products with a high level of healthy compounds. In this context, globe artichoke (*Cynara cardunculus* L. var. *scolymus* (L.) Fiori = *C. scolymus* L.), whose production for a long time was limited to the native Mediterranean region, is becoming increasingly popular and desired by consumers all over the world for its taste and functional properties [1–3]. The edible immature inflorescence (named as capitulum or head), which is composed of a receptacle surrounded by thickened bracts, is destined for fresh consumption or industrially processed production in several ways [4]. Particularly, the complex trimming operations necessary for heads to be freshly consumed, especially for the genotypes with spines, have increased recently the industrial production of ready-to-cook (RTC) globe artichoke slices [5,6], which can definitively balance consumer's demand due to easier use and freshness and better quality of them.

A good opportunity to improve sales for food industry is the production of fresh-cut vegetables; minimally processed products are one of the rapid growing sectors in the food industry due to the convenience and nutritional value, but it is well known that processing operations, mainly cutting, have the disadvantage of accelerating enzymatic browning phenomena and microbial spoilage, respect to the whole vegetables, due to the loss of compartmentalization of vegetable cells. Specifically, the enzymatic browning is mainly attributed to the polyphenol oxidase (PPO; EC 1.14.18.1), which generates dark pigments called melanoidins, unacceptable in terms of safety, since they could support the microbial growth. Additionally, although moderate, stress conditions or short storage times could be due to the loss of sensory and nutritional quality of the processed globe artichokes [5,7–11]. To slow down such quality decrease, the use of natural antimicrobials, such as essential oils (EOs), instead of chemical agents has emerged as a promising strategy [12]. Active packaging systems have been successfully applied to minimally processed vegetables [13]. Among them, active packaging releasing systems (emitters) are applied to add compounds into the headspace to improve shelf life of food, like antimicrobial substances, CO_2, antioxidants, or ethanol. Plants are relevant sources of bioactive molecules possessing antimicrobial activities against pathogens and spoilage microorganisms. The different compounds having antimicrobial activities in plant are alkaloids, phenolics, terpenes, terpenoids, flavonoids, essential oil, etc. [14,15]. The positive effects of the addition of EOs into individual carrier, as emitters inside packaging, instead of the incorporation into polymers, is to avoid any undesired change of properties of the polymeric films [16].

An important role in plant defense is played by EOs, secondary metabolites, some of which exert powerful antimicrobial activities [12,17,18], are classified as Generally Recognized As Safe (GRAS) [19], and are often used for the inhibition of pathogenic bacteria in foods.

EOs have been deeply studied also as active components in bio-based emulsified films and coatings. The EO from *Foeniculum vulgare* seeds, which is rich in trans-anethole, has been demonstrated to possess antioxidant activities [16], good antimicrobial activity against food-borne pathogens such as *Shigella dysenteriae* [20], *Escherichia coli* [21], and other pathogenic bacterial and fungal species [22–24]. With reference to food packaging application, recently Rizzo et al. [11] demonstrated the notable efficacy of the locust bean gum (LBG) coating with the addition of *F. vulgare* EO in preserving quality of RTC globe artichoke slices of "Spinoso sardo" during refrigerated storage for 11 days. However, although EO-based packaging has the ability to increase food shelf life, the selection of EO-based materials should be done on the basis of the compatibility with foods in terms of flavor to avoid any negative impact of EOs on the sensory profile.

The aims of this study, performed on RTC globe artichoke slices stored at 4 °C for 12 days, were therefore: (i) To evaluate the suitability of two globe artichoke genotypes, "Spinoso sardo" and "Opera F1", to be processed as RTC slices; (ii) to evaluate the effect of a *F. vulgare* EO emitter, within an active package system, as a strategy to slow down quality decay, and thus increasing shelf life; (iii) to estimate the impact of EO emitter within the package on the sensory profile of the RTC slices after microwave cooking.

2. Materials and Methods

2.1. Experimental Field, Plant Material, and Management Practices

Experimental trial was performed during 2014–2015 on Siracusa plain (36°58′ N, 15°11′ E; 53 m a.s.l. (metres above sea level)), one of the main Italian site for globe artichoke crop. The soil, classified as calcixerollic xerochrepts [25], had the following characteristics: pH 7.7, 50% sand, 18% silt, 32% clay, 6% limestone, 1.9% organic matter, 0.18% total N, and 30 mg kg^{-1} of available P_2O_5 and 280 mg kg^{-1} of exchangeable K_2O. The climate of the area is semiarid-Mediterranean, with mean long-term monthly maximum and minimum temperatures from 14.8 (January) to 30.6 °C (July) and from 7.8 (January) to 22.3 °C (August), respectively [26].

"Spinoso sardo", an early reflowering multiclone globe artichoke genotype, with conical green heads characterized by purple shades and big yellow spines [27], and "Opera F1", a hybrid producing purple-colored spherical heads, were studied. Planting was performed by either semi-dormant offshoots or seedling) in August 2014, by adopting a randomized block experimental design with 4 replications and a planting density of 1.0 plant m^{-2}. Crop management was carried out in agreement with the usual commercial practice of the cultivation area.

2.2. Head Harvest, Post-Harvest Treatments, and Sampling

Around 100 globe artichoke heads per each plot were harvested at the beginning of March at the marketable stage [28], transported to the Department of Agriculture, Food and Environment, University of Catania (Di3A) laboratories under controlled temperature and processed as a RTC product the day after. Samples were treated as reported by Licciardello et al. [9]. Briefly, all the inedible parts and the heads" tips were first eliminated and then heads were cut into 5 mm thick slices by using a manual cutting machine, sanitized and treated in anti-browning solution as reported by Rizzo et al. [11]. Slices were dried in a manual centrifuge, to eliminate residual water solution and then were placed in PET trays (16 × 11 × 3.5 cm), filled them up to 100 ± 10 g for each tray and packaged into a semi-permeable polyolefine film (SPP) (SP/BY 19 micron-System Packaging s.r.l., Siracusa, Italy) having an oxygen transpiration rate (OTR) of 3000 $cm^3/m^2/24$ h at 23 °C and 0% relative humidity (RH). Totally 80 trays were filled with "Opera F1" and "Spinoso sardo" slices respectively. Inside a half of the totally package for each genotype (40) was placed a small square (1 × 1 cm) of sterilized TNT textile, soaked with 150 μL of *F. vulgare* EO (0.75% *w*/*w*) produced by Rao Erbe (Catania, Italy) immediately before the hermetical sealing of trays with SPP plastic bags (20 × 15 cm) by a sealing bar (Lafayette Model SK-410, Frosinone, Italy). Samples were kept at 4 ± 0.5 °C and 90–95% RH up to 12 days and analysis were done at the processing day (T0), and after 5, 8 and 12 days of refrigerated storage (Figure S1).

2.3. Weight Loss and Headspace Gas Composition

Every package was marked with the starting weight. Then, at each sampling time three different trays for each post-harvest treatment were chosen (5, 8 and 12 day) and immediately weighed before further analysis. Mass loss (ML) was expressed as % of the initial sample weight at T0. The headspace gas composition, to quantify carbon dioxide and oxygen, was measured using a CheckPoint portable gas analyzer (MOCON Europe A/S (Dansensor), Ringsted, Denmark). Analysis were done on three replicates every sampling time.

2.4. Color Analysis

Surface color of RTC globe artichoke slices was assessed according to CIE L*a*b* scale as stated by Rizzo et al. [29], using a portable colorimeter (NR-3000, Nippon Denshoku Ind. Co., Ltd., Tokyo, Japan), correctly calibrated, and illuminant D65/10°. Data were showed as L* = lightness, a* = redness, b* = yellowness and calculated as $\Delta E^* = (\Delta L^{*2} + \Delta a^{*2} + \Delta b^{*2})^{1/2}$ pointing at the total color difference, to better assess the overall color changes during storage as previously done by Licciardello et al. [30].

2.5. Texture Analysis

The texture was measured by applying a 500 N nominal force cell and a stainless-steel probe (length 5 mm) and measuring the maximum shear force, using instruments and data elaboration as reported by Rizzo et al. [11] Results were the average of 6 measurements.

2.6. Microbiological Analyses

Microbiological determinations were performed on 25 g of RTC artichokes, aseptically sampled from each package and homogenized in a Lab-Blender 400 (Brinkmann, Westbury, NY, USA) for 3 min with 225 mL of Ringer solution (Oxoid, BR0052, Basingstoke, UK). The determined microorganisms were: Total aerobic mesophilic bacteria (MB) and total aerobic psychrotrophic bacteria (PB) on plate count agar (PCA, Oxoid, CM325, Basingstoke, UK) with cycloheximide 0.1% solution (Oxoid, SR0222, Basingstoke, UK), incubated at 30 °C for 48 h and at 7 °C for 5–10 days, respectively; at; yeasts and molds (YM) on Sabouraud Dextrose Agar (SDA, Oxoid, CM0041, Basingstoke, UK) supplemented with 0.1 g L^{-1} chloramphenicol (Oxoid, SR0078, Basingstoke, UK), incubated at 25 °C for 48–72 h; Enterobacteria (TEB) on violet red bile glucose agar (VRBGA, Oxoid, CM0485, Basingstoke,

UK), incubated at 37 °C for 24 h; *Escherichia coli* on brilliance E. coli selective agar (Oxoid, CM1046, Basingstoke, UK), incubated at 37 °C for 24 h; *Pseudomonas* spp. on Pseudomonas Agar Base (CM0559, Oxoid, Basingstoke, UK), supplemented with Pseudomonas CFC selective agar supplement (SR0103, Oxoid, Basingstoke, UK) and incubated at 25 °C for 48 h. All microbiological analyses were performed in triplicate and expressed as average log10 CFU g^{-1}.

2.7. Chemical and Enzymatic Analyses

An amount of RTC globe artichoke slices from each package was freeze dried (Christ freeze drier; Christ, Osterode am Harz, Germany) for performing the following chemical determinations.

L-ascorbic acid content (AsAC) was evaluated following the HPLC method proposed by Lombardo et al. [5] and expressed as mg kg^{-1} of DM (dry matter).

A modified Folin-Ciocalteu method [4] was adopted for the determination of the total polyphenol content (TPC), which was expressed as g kg^{-1} of DM, using chlorogenic acid as a standard; in the same extracts used for TPC analysis, the antioxidant activity (AA), expressed in terms of DPPH (2,2-diphenyl-1-picrylhydrazyl) percentage of inhibition, was determined using the method reported by Brand-Williams et al. [31] and calculating the results as follows:

$$AA = (AC0 - AS30) \div AC0 \times 100 \quad (1)$$

where AC0 is the absorbance of the control assay (no extract) and AS30 the absorbance of the sample after 30 min.

The determination of PPO activity was spectrophotometrically performed [32,33], using catechol as a phenolic substrate. defining one unit of PPO activity as the level of enzyme able to increase the absorbance by 0.001 min^{-1}.

All these chemical determinations were conducted by using bi-distilled water and analytical or HPLC grade reagents and solvents obtained from Sigma-Aldrich (Milan, Italy).

2.8. Sensory Analyses

The UNI EN ISO 13,299 [34] method was adopted to evaluate changes in sensory characteristics of samples. Ten judges (six female and four male, 24–40 years old and a plurennial expertise in the sensory evaluation of vegetables) were selected from the Di3A and were trained as indicated by ISO 8586 [35] in four meetings using different kind of globe artichoke samples, to achieve a common language for the description of sensory traits and to familiarize themselves with scales and procedures.

According to Rizzo et al. [11,17] the following main sensory attributes were considered: One for appearance (freshness), six for odor (herbaceous, globe artichoke, fennel, apple, potato, and off-odor), six for flavor (herbaceous, globe artichoke, fennel, apple, potato, and off-flavor), two for taste (sweet and bitter), one for tactile (astringent), and overall score. Before the sensory evaluation, the samples were cooked using commercially available microwave cooking bags (FRIO; Sphere France S.A.S., Paris, France) in a microwave oven (Whirlpool, Benton Harbor, MI, USA) at 700 W for 10 min.

The panel evaluated the randomized samples adopting a scale from 1 (absence of sensation) to 9 (extremely intense) in the sensory laboratory [36] of the Di3A. Water was provided for mouth rinsing between tests. Computerized data-collection software was used (FIZZ, Software Solutions for Sensory Analysis and Consumer Tests, Biosystemes, Couternon, France).

2.9. Statistical Analysis

Bartlett's test was adopted to verify the homoscedasticity, and then the data (when necessary transformed by Bliss transformation prior to statistical analysis) were subjected to a three-way analysis of variance (ANOVA) as a factorial combination of "genotype (2) × postharvest treatment (2) × storage time (4)". For each sensory attribute a two-way ANOVA as a factorial combination of "genotype (2) × postharvest treatment (2)" was performed. Means were separated by LSD test, when the *F*-test was significant ($p \leq 0.05$). Limitedly to AsAC, TPC, and AA, the correlation analysis was carried out to estimate their relationship. The software package Statgraphics® CenturionXVI (Statpoint Technologies, INC., The Plains, VA, USA) was used to perform ANOVA.

3. Results and Discussion

3.1. Fresh Weight Loss, Headspace Concentration, Color, and Texture

It is assumed that fresh globe artichoke heads are cooked prior to consumption. The USDA National Nutrient Database for Standard Reference reports that about 40% of a fresh head is edible, while the USDA Food and Nutrient Database for Dietary Studies indicates that about 6% of this edible share is further lost through cooking. Therefore, while the consumers have to consider only the weight lost through cooking of RTC globe artichoke slices, producers have to consider also the whole mass loss to define their profit. As previously reported by Licciardello et al. [9], the most important function of the packaging is to decrease the fresh vegetable moisture loss by respiration and surface evaporation mechanisms, thus increasing product shelf life. For this reason, the SPP film was selected for our study, considering the positive performance in reducing fresh weight loss under refrigerated storage conditions at a maximum−1%, due to its low water vapor transmission rates [6]. The ANOVA for the ML highlighted a significant influence of the studied main effects (Genotype (G), Treatment (T) and Storage time (St)), as well as a high statistical significance for the interaction "G × St", "T × St" and "G × T × St" (Table 1). As expected, ML increased during the refrigerated storage (Table 2), reaching its maximum value around 2%, which resulted quite higher than the observed values for other globe artichoke genotypes previously studied (i.e., "Apollo", "Exploter", and "Spiosodi Palermo") [9].

ML was slightly higher in "Opera F1" at the end of the experimental trial (12 day) respect to "Spinoso sardo", while samples with EO"s pad began with higher values that were maintained until 8 days (Figure 1). It is evident how the RTC globe artichoke slices packed with EO emitter follows the same path of the control, underlining more correspondence to the genotype then to the treatment (Figure 1).

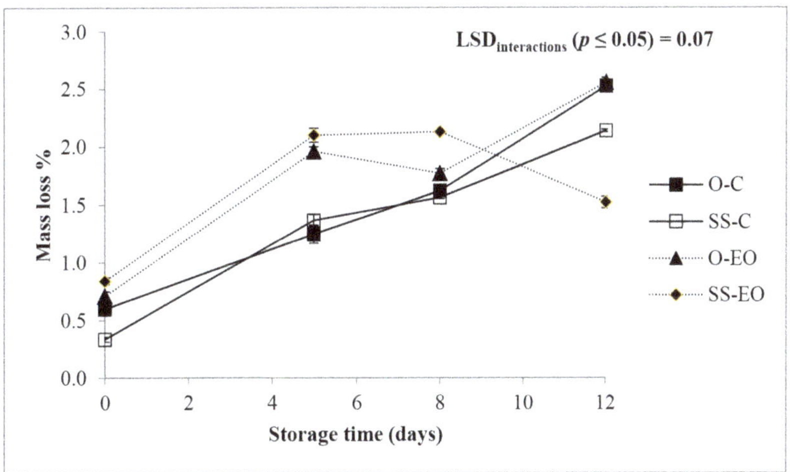

Figure 1. Mass loss (%) of RTC globe artichoke slices as affected by "genotype × post-harvest treatment × storage time" interaction. O: "Opera F1"; SS: "Spinoso sardo"; C: Control; EO: Samples packed with essential oil of *Foeniculum vulgare*. Data are reported as means ± standard deviation.

Table 1. Mean square as absolute value and percentage of total (in brackets) of effects resulting from analysis of variance.

(a)

		Qualitative Trait										
Source of Variation	**df**	**ML%**	**O_2%**	**CO_2%**	**TPC**	**AsAC**	**AA**	**Text**	**L***	**a***	**b***	**ΔE**
Genotype (G)	1	0.13 *** (2.46)	4.20 ^{NS} (3.02)	4.10 ^{NS} (5.84)	43.01 * (4.75)	2485 *** (15.24)	304 *** (38.07)	347.49 * (20.02)	0.03 ^{NS} (0.3)	1.87 ^{NS} (1.43)	0.04 ^{NS} (0)	0.81 ^{NS} (1.4)
Treatment (T)	1	0.6 *** (11.82)	0.0 ^{NS} (0)	20.11 * (28.66)	118.20 ** (13.05)	1300 *** (7.98)	112 *** (14.10)	89.28 ^{NS} (5.14)	1.19 ** (6.2)	5.74 ^{NS} (4.4)	1.06 ^{NS} (2.3)	2.11 ^{NS} (3.64)
Storage time (St)	3	3.54 *** (69.75)	80.56 *** (57.95)	30.31 ** (43.2)	698.14 *** (77.10)	11,577 *** (71)	321 *** (40.25)	113.75 ^{NS} (6.55)	7.35 *** (71.78)	104.1 ** (79.79)	37.54 *** (81.38)	45.41 *** (78.36)
(G) × (T)	1	0.0 ^{NS} (0)	10.89 ^{NS} (7.83)	2.28 ^{NS} (3.25)	1.58 ^{NS} (0.17)	4.5 ^{NS} (0.03)	0.08 ^{NS} (0.01)	10.27 ^{NS} (0.6)	0.49 ^{NS} (4.79)	0.31 ^{NS} (0.24)	0.11 ^{NS} (0.24)	0.77 ^{NS} (1.33)
(G) × (St)	3	0.33 *** (6.5)	15.90 * (11.44)	3.32 ^{NS} (4.73)	33.23 * (3.67)	118 ^{NS} (0.72)	32 *** (4.04)	333.91 * (19.24)	0.14 ^{NS} (1.3)	7.51 ^{NS} (5.8)	2.88 ^{NS} (6.2)	2.11 * (3.64)
(T) × (St)	3	0.36 *** (7.09)	16.26 * (11.70)	6.91 ^{NS} (9.85)	6.45 * (0.71)	815 *** (5)	21 *** (2.64)	338.91 * (19.52)	0.09 ^{NS} (0.9)	1.04 ^{NS} (0.8)	0.40 ^{NS} (0.9)	6.12 *** (10.56)
(G) × (T) × (St)	3	0.12 *** (2.36)	11.21 ^{NS} (8.06)	3.13 ^{NS} (4.46)	4.89 ^{NS} (0.54)	5.2 ^{NS} (0.03)	7 ^{NS} (0.89)	502.20 ** (28.93)	0.95 ** (9.28)	9.89 ^{NS} (7.58)	4.10 ^{NS} (8.9)	0.62 ^{NS} (1.1)
Total mean square		5.07	139.02	70.16	905.49	16,306	798	1735.81	10.24	130.46	46.13	57.95

(b)

		Qualitative Trait						
Source of Variation	**df**	**MB**	**PB**	**YM**	**Enterobacteria**	***Pseudomonas***	***E. coli***	**PPO**
Genotype (G)	1	11.59 *** (55)	17.66 *** (61.71)	33.84 *** (58.5)	4.69 *** (24.41)	4.13 *** (10.31)	2.12 *** (4.35)	4.14 *** (22.73)
Treatment (T)	1	1.06 *** (5)	1.06 *** (3.7)	0.77 *** (1.3)	0.49 *** (2.55)	1.75 *** (4.37)	7.51 *** (15.42)	1.58 *** (8.68)
Storage time (St)	3	7.74 *** (36.73)	9.49 *** (33.16)	22.70 *** (39.23)	13.46 *** (70.07)	30.07 *** (75.04)	26.70 *** (54.81)	9.19 *** (50.47)
(G) × (T)	1	8×10^{-4} ^{NS} (0)	0.02 *** (0)	0.03 *** (0)	0.07 *** (0.36)	0.29 *** (0.72)	1.07 *** (2.2)	0.09 ^{NS} (0.5)
(G) × (St)	3	0.53 *** (2.5)	0.22 *** (0.7)	0.40 *** (0.7)	0.37 *** (1.93)	3.49 *** (8.71)	8.46 *** (17.37)	2.07 *** (11.37)
(T) × (St)	3	0.12 *** (0.6)	0.13 *** (0.45)	0.09 *** (0.16)	0.10 *** (0.5)	0.16 *** (0.4)	2.22 *** (4.56)	1.12 *** (6.15)
(G) × (T) × (St)	3	0.03 *** (0.14)	0.04 *** (0.14)	0.03 *** (0)	0.03 *** (0.16)	0.18 *** (0.45)	0.63 *** (1.29)	0.02 ^{NS} (0.1)
Total mean square		21.07	28.62	57.86	19.21	40.07	48.71	18.21

Note: ML: Mass loss; TPC: Total polyphenol content; AsAC: Ascorbic acid content; AA: Antioxidant activity; Text: Texture (F_{max}); color parameters (L*, a*, b*); ΔE: total color difference; df: Degrees of freedom. MB: Mesophilic bacteria; PB: Psychrotrophic bacteria; YM: Yeasts and molds; PPO: Polyphenol oxidase; *, ** and *** indicate significant at $p \leq 0.05$, $p \leq 0.01$ and $p \leq 0.001$, and ^{NS}, not significant.

Table 2. Qualitative traits of ready-to-cook (RTC) globe artichoke slices as affected by the main factors under study.

	Main Factor	Qualitative Trait										
		ML%	O_2%	CO_2%	TPC (g kg^{-1} DM)	AsAC (mg kg^{-1} DM)	AA (% DPPH inhibition)	Text (N)	L*	a*	b*	ΔE
							Genotype					
(a)	*OPERA F1*	1.62 [a]	4.57 [a]	12.26 [a]	21.26 [a]	118 [a]	56.6 [a]	63.83 [a]	92.80 [a]	2.17 [a]	0.31 [a]	2.79 [a]
	SPINOSO SARDO	1.49 [b]	5.29 [a]	12.98 [a]	18.94 [b]	100 [b]	50.4 [b]	57.24 [b]	92.87 [a]	2.65 [a]	0.24 [a]	3.11 [a]
							Treatment					
	Control	1.42 [b]	4.94 [a]	13.41 [a]	18.18 [b]	103 [b]	51.6 [b]	62.20 [a]	93.03 [a]	1.98 [a]	0.46 [a]	2.69 [a]
	Essential oil	1.69 [a]	4.92 [a]	11.83 [b]	22.02 [a]	116 [a]	55.4 [a]	58.86 [a]	92.65 [b]	2.83 [a]	0.09 [a]	3.21 [a]
							Storage time (day)					
	0	0.62 [d]	9.24 [a]	11.36 [b]	31.6 [a]	165 [a]	61.7 [a]	64.82 [a]	93.26 [b]	3.43 [a]	−0.53 [b]	-
	5	1.56 [c]	2.14 [c]	14.77 [a]	23.78 [b]	102 [b]	55.3 [b]	55.80 [a]	92.03 [c]	4.47 [a]	−0.92 [b]	3.65 [b]
	8	1.77 [b]	3.02 [b,c]	13.77 [a]	13.15 [c]	90 [c]	48.8 [c]	59.79 [a]	93.99 [a]	−2.93 [b]	3.51 [a]	5.70 [a]
	12	2.18 [a]	5.33 [b]	10.61 [b]	11.87 [c]	80 [d]	48.2 [c]	61.71 [a]	92.07 [c]	4.68 [a]	−0.95 [b]	2.46 [c]

	Main Factor	Qualitative Trait						
		MB	PB	YM	Enterobacteria	*E. coli*	*Pseudomonas*	PPO
		(log CFU g^{-1})						(mmol min^{-1} g^{-1})
					Genotype			
(b)	*OPERA F1*	3.89 [b]	3.64 [b]	3.13 [b]	4.53 [b]	1.64 [b]	4.12 [b]	2.65 [b]
	SPINOSO SARDO	5.09 [a]	5.13 [a]	5.18 [a]	5.30 [a]	2.23 [a]	4.84 [a]	3.37 [a]
					Treatment			
	Control	4.68 [a]	4.57 [a]	4.31 [a]	5.04 [a]	2.15 [a]	4.72 [a]	3.23 [a]
	Essential oil	4.31 [b]	4.20 [b]	4 [b]	4.79 [b]	1.72 [b]	4.25 [b]	2.79 [b]
					Storage time (day)			
	0	3.21 [d]	3.23 [d]	1.8 [d]	3.07 [d]	0.8 [d]	1.63 [c]	2.01 [b]
	5	4.33 [c]	3.99 [c]	4.08 [c]	4.96 [c]	1.68 [c]	4.93 [b]	4.03 [a]
	8	4.92 [b]	4.51 [b]	5.28 [b]	5.70 [b]	2.35 [b]	5.70 [a]	3.83 [a]
	12	5.52 [a]	5.82 [a]	5.46 [a]	5.92 [a]	2.91 [a]	5.68 [a]	2.16 [b]

Note: ML: Mass loss; TPC: Total polyphenol content; AsAC: Ascorbic acid content; AA: Antioxidant activity; Text: Texture (F_{max}); color parameters (L*, a*, b*); ΔE: total color difference. MB: Mesophilic bacteria; PB: Psychrotrophic bacteria; YM: Yeasts and molds; PPO: Polyphenol oxidase. Different letters within the same qualitative trait and main factor show significant differences (LSD test, $p \leq 0.05$).

With reference to the two components of packaging headspace, both O_2 and CO_2 were not influenced by G, as expected, thus confirming previous results [9], while St influenced the percentage of O_2 and CO_2 for $p \leq 0.001$ and $p \leq 0.01$, respectively (Table 1). Changes in O_2 and CO_2 concentration slowed down during the cold storage (Table 2), while only the O_2 concentration varied during the storage time depending on G and T as reported in Figure 2a,b, respectively. Thus, the headspace O_2 concentration changed rapidly during the first 5 days of storage [5,9], then reaching an equilibrium concentration until the end of the storage for both G and T. "Opera F1" displayed higher O_2 concentration than "Spinoso sardo" only at the processing day, while both genotypes reported similar values of this parameter throughout the storage time (Figure 2a). The presence of EO emitter instead seemed to slow down the consumption of O_2, reaching after 12 days comparable headspace gas concentrations (Figure 2b). As explained by Ghidelli et al. [37], preparing globe artichoke heads by removing inedible parts led to a sharp drop in O_2 and an increase in CO_2.

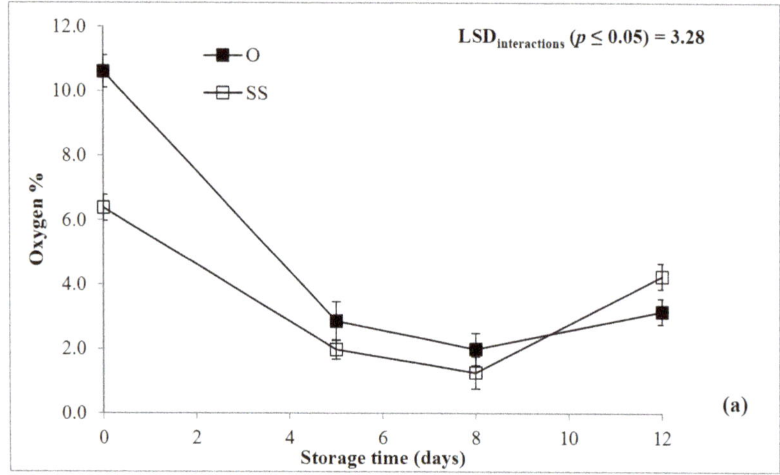

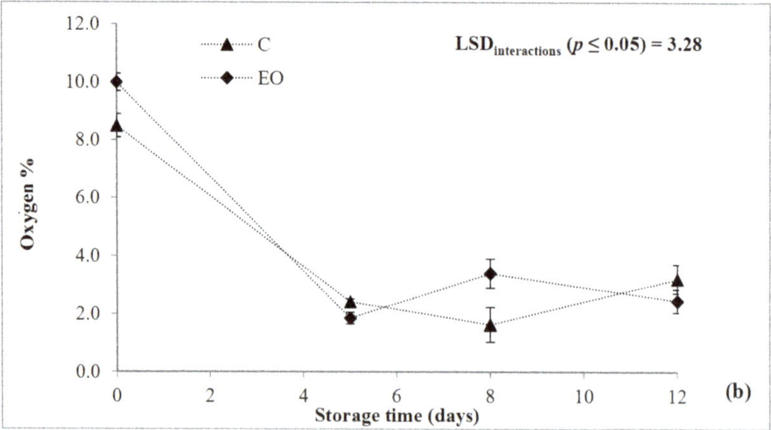

Figure 2. Oxygen (%) of RTC globe artichoke slices as affected by "genotype × storage time" (**a**); "post-harvest treatment × storage time" (**b**); interaction. O: "Opera F1"; SS: "Spinoso sardo"; C: control; essential oil (EO): Samples packed with essential oil of *F. vulgare*. Data are reported as means ± standard deviation.

In fresh cut vegetables, appearance color represents the marketability value of the product itself. Considering the studied chromatic indexes, the ANOVA showed as all the color parameters were mainly affected by St (Table 1) in agreement with previous findings, which reported as St resulted the predominant factor influencing color [10]. L* was affected by T, St and "G × T × St" interaction (Table 1; Figure 3a). L* decreased in both genotypes and post-harvest treatments during the first 5 days, as confirmed by previous study [11], then increased in the following 3 days and again slowed down at the end of the trial. "Opera F1" had higher values from T0 to the 8th day of St, to rapidly decrease at the last day of the trial respect to "Spinoso sardo", which instead had higher values at the end of St. Such path may be explained considering that studying the color by a colorimeter, a decrease of L* can be an indication of browning appearance as well as an increase in L* can be linked with the progress of whiteness in the samples for the same reason [38].

The ΔE was characterized by "G × St" and "T × St" interactions (Table 1). In this view, the genotype "Spinoso sardo" recorded higher values than "Opera F1", thus indicating greater changes during the storage period (Figure 3b), and the presence of EO's pad did not improve color appearance, since the increase of ΔE after 8 days corresponds to a worse color vision by the consumer (Figure 3c), but at the end of the storage it was smaller than control. Generally, ΔE assessed color changes taking place during St, which is not simply related to browning or to other perceived aspects, but it is the expression of the complexity of different (unspecified) effects [30].

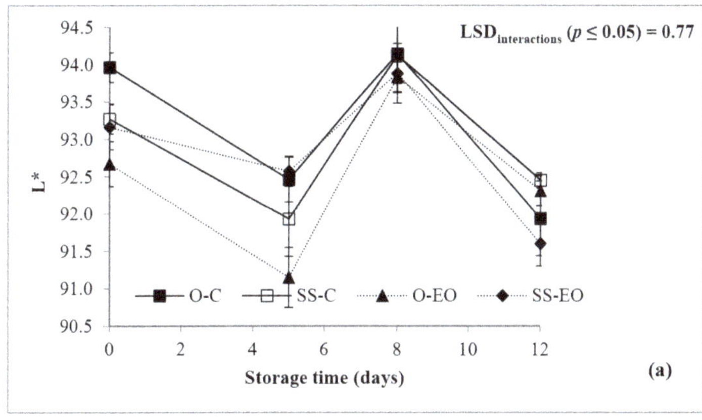

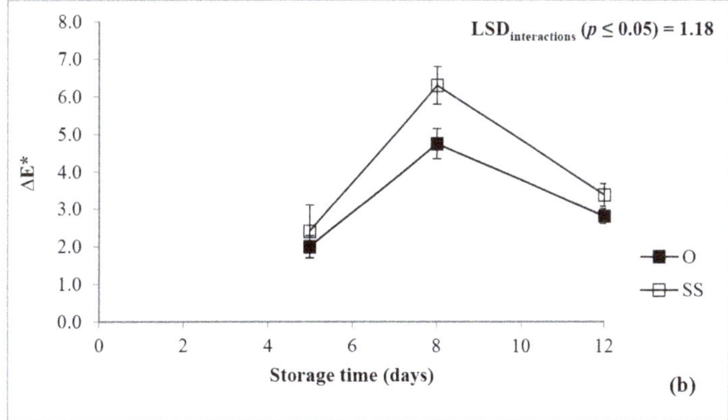

Figure 3. *Cont.*

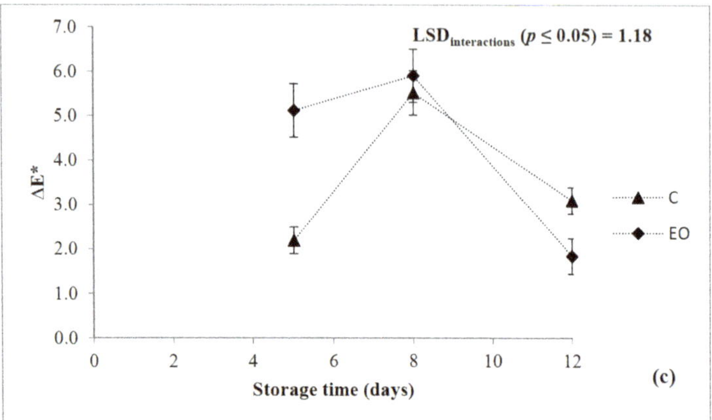

Figure 3. Color parameters of RTC globe artichoke slices as affected by "genotype × post-harvest treatment × storage time"(**a**) "genotype × storage time" (**b**); "post-harvest treatment × storage time"(**c**) interaction. L*: Lightness; ΔE* = total color difference; O: "Opera F1"; SS: "Spinoso sardo"; C: control; EO: Samples packed with essential oil of *F. vulgare*. Data are presented as means ± standard deviation.

In RTC globe artichoke slices texture is doubly important because it has to satisfy the consumer at package's opening, but also after cooking. Texture was significantly influenced by interactions "G × St", "T × St" (both at $p \leq 0.05$) and "G × T × St" ($p \leq 0.01$), as reported in Table 1. Texture in "Opera F1" did not change substantially throughout the storage time, while "Spinoso sardo" reported an immediate loss of about 50% after the first 5 days, then settling on almost constant but lower values than those recorded for "Opera F1" (Figure 4). As studied in the past, texture increased in "Opera F1" for the loss of water by the vegetable, possibly derived by lignin production, while the resistance decreased in "Spinoso sardo" during storage for the liberation of proteolytic and pectolytic enzymes caused by the cellular breakdown [11]. Also, the EO treatment showed a fairly increasing trend, keeping values after 5 days of storage slightly higher than the control which showed a drop off after 5 days (Figure 4).

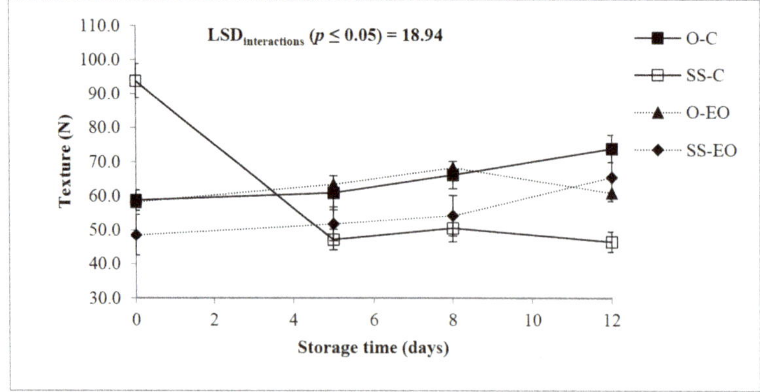

Figure 4. Texture (N) of RTC globe artichoke slices as affected by "genotype × post-harvest treatment × storage time" interaction. O: "Opera F1"; SS: "Spinoso sardo"; C: Control; EO: Samples packed with essential oil of *F. vulgare*. Data are showed as means ± standard deviation.

3.2. Antioxidant Compounds Content and Activity

Inhibition of browning phenomena is a main challenge in the production of RTC globe artichoke slices with the aim of ensuring a longer shelf life and more suitable visual qualitative attributes of the product [9,10]. Natural antioxidant compounds, either endogenous in globe artichoke (e.g., ascorbic acid and polyphenols) or applied as safe preservatives (e.g., EOs), are found to be efficient in preventing the qualitative damages during cold storage of minimally processed vegetables [6,30]. However, several factors influenced the levels of antioxidant compounds in globe artichoke [39–41]. Here, the ANOVA results revealed the significant influence of the studied main effects (G, T and St) and the "T × St" interaction for both the AsAC and TPC (Table 1).

Both TPC and AsAC decreased throughout the storage time, regardless of post-harvest treatment and genotype (Table 2). For the TPC, this may depend upon the conversion of phenolic compounds to the relative quinones (scarcely reactive to the Folin–Ciocalteu method) by PPO [42]. For such reason, when the PPO activity declined, the TPC was quite stable in both post-harvest treatments (Figure 5) and genotypes (Figure 6) passing from 8 to 12 days of cold storage. Our results also indicated that TPC decreased much more slowly in presence of EO (Figure 5), which can be imputed to the additional polyphenols provided by packing RTC globe artichoke slices with EO emitter. In particular, after 5 and 8 days of cold storage, EO was able to increase the TPC than the control C (Figure 5). Similarly, during the storage time, samples presented a different AsAC retention with reference to the post-harvest treatment subjected (Figure 5). In particular, after 5 days of cold storage EO reinforced the AsAC retention as also highlighted for the TPC.

The AA (measured by DPPH assay) had a good correlation with both TPC (r = 0.821 ***) and AsAC (r = 0.775 ***). As reported previously [6,10], globe artichoke extracts possess a high and exploitable AA as safe qualitative preservative. Here, it was significantly influenced by the studied main effects (G, T, and St) and the interactions "T × St" and "G × St" (Table 1). On the whole, the AA trend is consistent with those highlighted for AsAC and TPC, decreasing throughout the storage period, although with a different extent depending on post-harvest treatment (Figure 5) and genotype (Figure 6). Indeed, at each sampling time it was higher in samples packed with EO emitter than C (Figure 5), as a result of the enhanced level of antioxidants in the product provided by EO of *F. vulgare*.

A significant influence of genotype was found for all the three traits here commented (Table 1). In this view, the decoding of globe artichoke genome may lead to the constitution of genotypes with enhanced yield and quality features [3,43,44]. Here, it is noteworthy to underline as "Opera F1" displayed higher TPC and AsAC than "Spinoso sardo", as well as a higher AA consequently (Table 2). However, "Opera F1" experienced a higher reduction of AA at each sampling time (Figure 6), which could be related to its polyphenolic profile. Indeed, polyphenols are recognized as the main contributors to the AA of globe artichoke extracts [1,10].

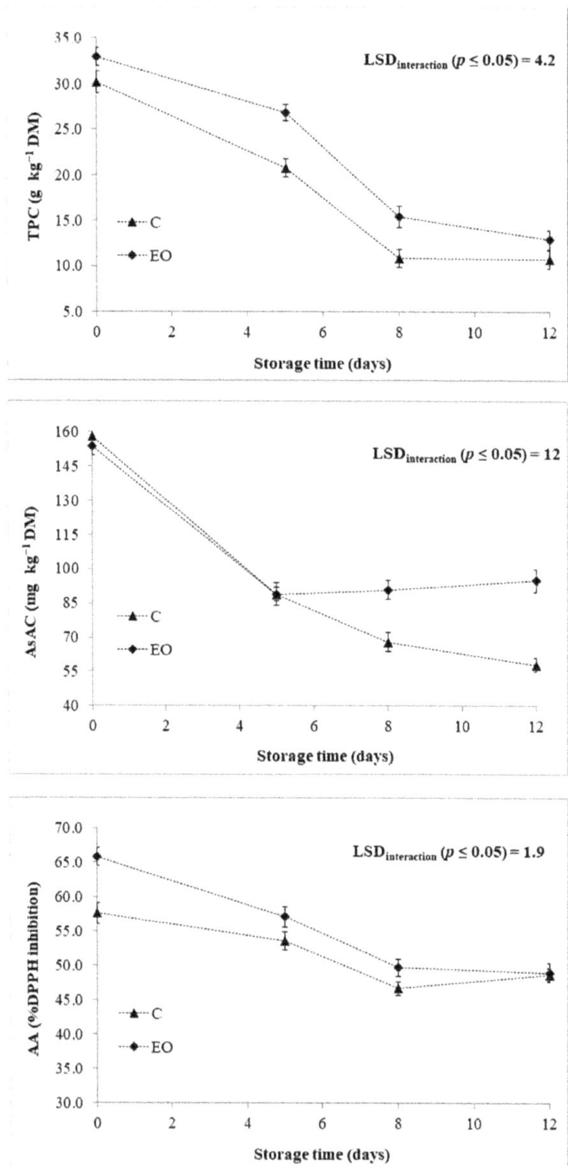

Figure 5. Chemical traits of RTC globe artichoke slices as affected by "post-harvest treatment × storage time" interaction. TPC: Total polyphenol content; AsAC: Ascorbic acid content; AA: Antioxidant activity; C: Control; EO: Samples packed with essential oil of *F. vulgare*. Data are presented as means ± standard deviation.

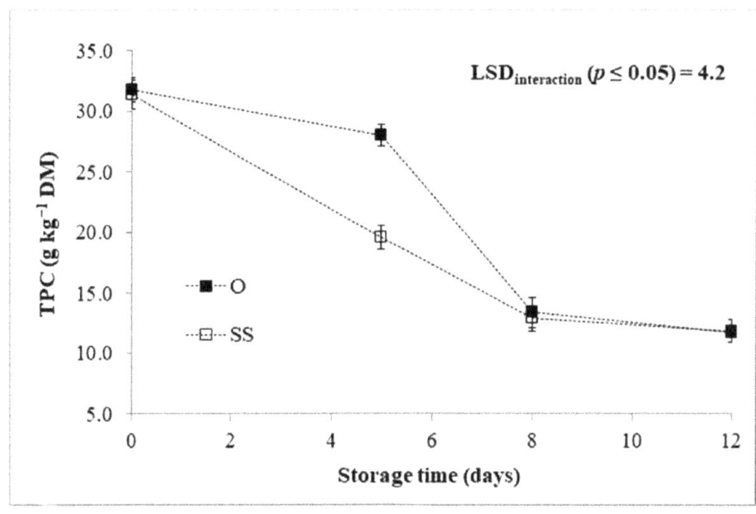

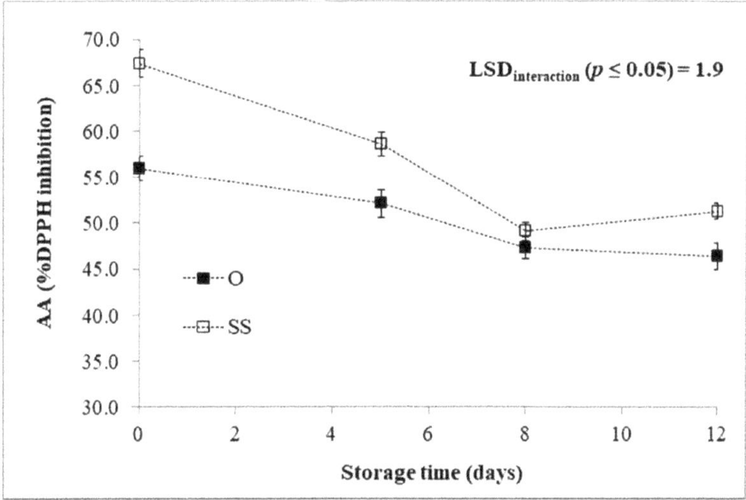

Figure 6. Chemical traits of RTC globe artichoke slices as affected by "genotype × storage time" interaction. TPC: Total polyphenol content; AA: Antioxidant activity; O: "Opera F1"; SS: "Spinoso sardo". Data are presented as means ± standard deviation.

3.3. Polyphenol Oxidase (PPO) Activity

The ANOVA revealed that the three sources of variation (G, T and S) were all individually significant for the PPO parameter, but only "G × St" and "T × St" interactions were significant (Table 1).

Based on the significance of the "G × St" interaction, a similar trend was observed for the PPO in both the selected genotypes (Figure 7a), although the concentration of this enzyme was in the early stages of storage time significantly higher in "Spinoso sardo" compared to "Opera F1", making the latter genotype potentially the best choice for industrial transformation. The PPO enzyme is activated by stress conditions, resulting after least operations of mincing and cutting and/or by premature ageing phenomena. The genotype "Opera F1" showed at the day of processing (T0), a lower endogenous PPO activity (1.65 mmol min^{-1} g^{-1}) respect to "Spinoso sardo" (2.36 mmol min^{-1} g^{-1}). In addition, "Spinoso sardo" shows a maximum of activity after the 5th day of storage, while in "Opera F1" it appeared delayed at 8 days, but lower than "Spinoso sardo" in the same sampling time. This tendency should be linked with the browning reactions activated by specific polyphenolic substrates. As previously observed, the extent of the peroxidase (POD, EC 1.11.1.7) isoenzymes in "Opera F1" is really poorer if compared to other globe artichoke genotypes or other crops, such as tomato, melon and strawberry [45]. In addition, variations in polyphenols, PPO and PAL activities were previously reported in cold-stored globe artichoke heads [46]. The subsequent reduction in enzymatic activity may be ascribable to the reduction of the endogenous PPO, which is responsible of the browning phenomena [46], as well as to a molecular rearrangement of chlorogenic acid, 1,3-*O*- and 3,5-*O*-dicaffeoylquinic acids, which are the most representative phenolic compounds in this crop [11,47,48].

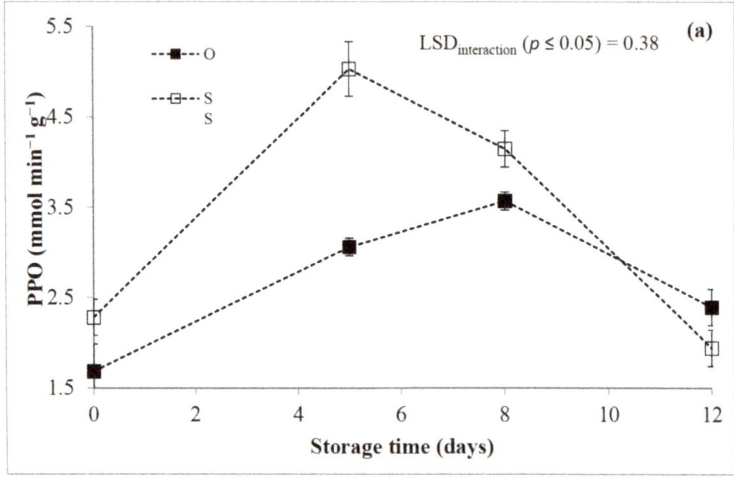

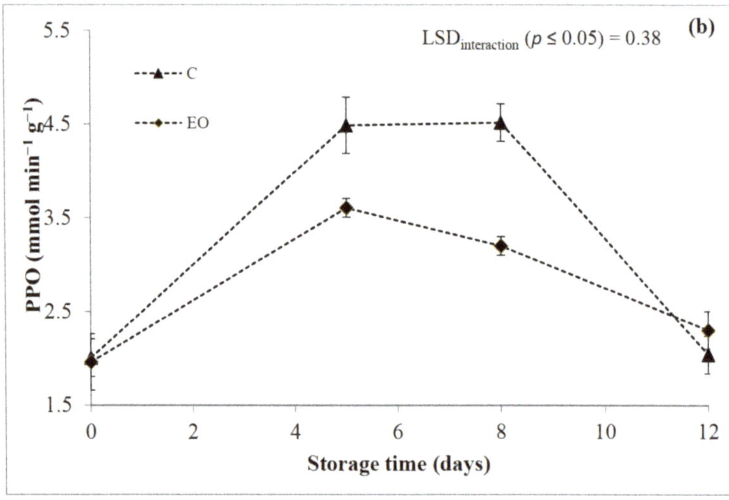

Figure 7. PPO activity of RTC globe artichoke slices as affected by "genotype × storage time" (**a**) and "treatment × storage time" (**b**) interaction. PPO: Polyphenol oxidase; O: "Opera F1"; SS: "Spinoso sardo".C: Control; EO: Samples packed with essential oil of *F. vulgare*. Data are presented as means ± standard deviation.

Referring to the interaction "T × St" (Figure 7b), both the control (C) and EO treatment had a similar pattern, showing an increase of the values up to the 5th day of refrigerated storage, followed by a gradual decrease at 8 days in EO, confirming the effectiveness of the treatment in reducing the incidence of the polymerization of quinones in brown melanoidins than the control C. It should be noted that the initial browning phenomena could be also attributed to an increased availability of PPO due to its solubilization from the cell wall, mediated mainly by polygalacturonase (EC 3.2.1.15), and pectin methyl esterase (EC 3.1.1.11) [7,10].

3.4. Microbiological Quality

Table 1 shows that each of the three considered main factors (G, T, and St) significantly affected the studied microbiological parameters of RTC globe artichoke slices ($p \leq 0.001$). Regarding G, higher average microbial counts were recorded for "Spinoso sardo" than "Opera F1" (Table 2), probably for the fact that "Spinoso sardo" is characterized by less assurgent leaves and shorter flowering stems thus exposing heads to a more pronounced microbial contamination from the environment. Referring to St, all microbial groups gradually increased throughout the storage period (Table 2, Figure 8). In particular, in none of the sampling times PB and MB overcame the limit of 8 log CFU g^{-1} suggested by CNERNA-CNRS [49], of 7.7 log CFU g^{-1} imposed by the French law for fresh-cut vegetables [50] and of 7 log CFU g^{-1} as maximum value at expiry date dictated by the Spanish regulation for prepared meals [51]. Instead, the average YM levels at 8 and 12 day (6.03 log CFU g^{-1}) moderately exceeded the relative limit of 5 log CFU g^{-1} recommended by CNERNA-CNRS [49], but not causing sensory defects. The average Enterobacteria over the storage period was higher than that reported by Licciardello et al. [9], being influenced by the higher counts evidenced for "Spinoso sardo" already starting from the processing day; however, the average *E. coli* plate count was 2.91 log CFU g^{-1} at the end of the considered period, in conformity with European Regulation EC 1441/2007 [52]. Regarding T, when EO emitter was added to the package system, an average decrease of about 0.50 log CFU g^{-1} was observed for all microbial groups. As reported in Table 2, the "G × T × St" interaction significantly influenced all the considered microorganisms, which were significantly higher in control RTC slices than in those packed with EO emitter (Table 2, Figure 8). The greatest decrements were noticed at 12 d for MB, YM and Enterobacteria in "Spinoso sardo" (0.72, 0.62, and 0.5 log CFU g^{-1}, respectively) and for *Pseudomonas* spp. in "Opera F1" samples (0.76 log CFU g^{-1}).

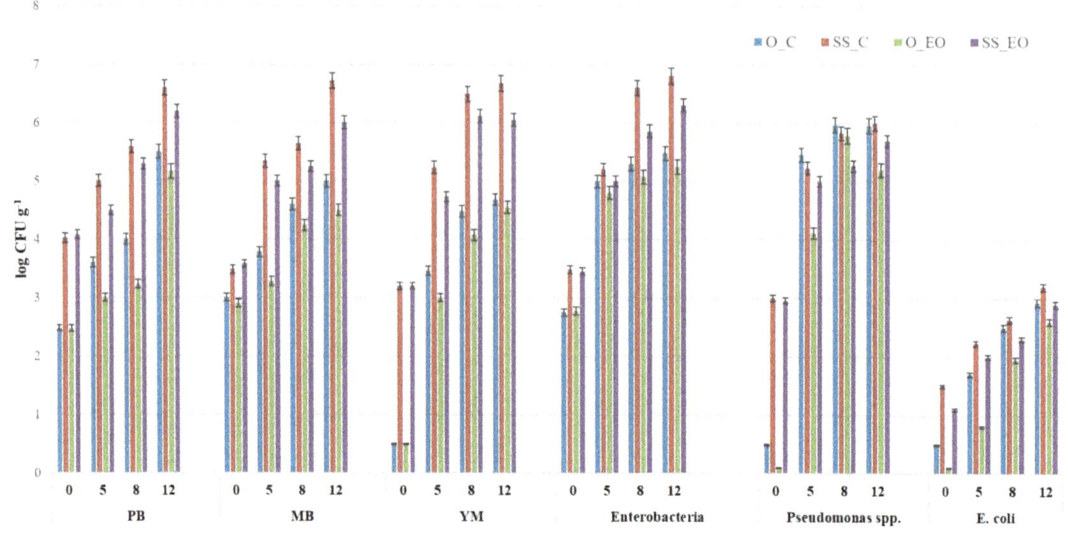

Figure 8. Microbiological parameters of RTC globe artichoke slices as affected by "genotype × post-harvest treatment × storage time" interaction. $LSD_{interaction}$ ($p \leq 0.05$): 0.06 (total mesophilic bacteria, MB); 0.07 (total psychrotrophic bacteria, PB); 0.05 (yeasts and molds, YM); 0.07 (Enterobacteria, TEB); 0.08 (*Pseudomonas* spp.); 0.08 (*E. coli*). Data are presented as means ± standard deviation.

3.5. Sensory Analysis

The sensory attributes that significantly differentiated, at each sampling time, the RTC globe artichoke slices of the two studied genotypes were shown in Table 3. The intensity (expressed as mean score) was reported only for the significantly different attributes.

Table 3. Mean scores of the significant sensory attributes of RTC globe artichoke slices as affected by genotype and post-harvest treatment under study.

Storage Time (d)	Attribute	Treatment			
		C		EO	
		"Opera F1"	"Spinoso sardo"	"Opera F1"	"Spinoso sardo"
0	Fennel odor	1.0 [a,1]	1.0 [a]	3.2 [b]	3.4 [b]
	Fennel flavor	1.0 [a]	1.0 [a]	3.4 [b]	3.3 [b]
5	Fennel odor	1.0 [a]	1.0 [a]	3.6 [b]	4.5 [b]
	Bitter	4.8 [b]	2.7 [a]	3.2 [ab]	2.5 [a]
	Fennel flavor	1.0 [a]	1.0 [a]	3.9 [b]	4.3 [b]
8	Fennel odor	1.0 [a]	1.0 [a]	2.6 [b]	4.5 [c]
	Fennel flavor	1.0 [a]	1.0 [a]	3.2 [b]	5.1 [c]
12	Fennel odor	1.0 [a]	1.0 [a]	3.6 [b]	3.3 [b]
	Fennel flavor	1.0 [a]	1.0 [a]	3.9 [b]	3.8 [b]

[1] Values marked with different letters in the same row are significantly different ($p \leq 0.05$) according to the LSD test. C: Control; EO: Samples packed with essential oil of *F. vulgare*.

At 0, 8, and 12 days, samples were perceived as similar by panelists, except for fennel odor and flavor, while after 5 days of refrigerated storage the samples were significantly different also for bitter attribute. The latter was more intense in the control C as especially observed for "Opera F1". As expected, the samples packed with EO emitter showed the highest intensities of fennel odor and flavor, regardless of genotype (Table 3).

4. Conclusions

The present findings revealed that both globe artichoke genotypes possess a good attitude to be processed as RTC slices, although with a difference response to the applied technological conditions: "Opera F1" showed the best performances for color appearance, thanks to the lower isoenzymes activity, texture and chemical indexes (i.e., TPC, AsAC and AA), while "Spinoso sardo" showed lower ML over the storage time. The addition of *F. vulgare* EO emitter into the packaging system did not influence ML, color appearance and texture, while its presence slowed down the consumption of O_2, better preserved texture when compared to the control and more effectively controlled PPO activity and antioxidants (AsAC and TPC) retention during the cold storage.

As regards the microbiological quality, all the considered microbial groups gradually increased until the end of the storage period, with a significant higher extent in control RTC globe artichoke slices than in those packed with EO emitter, confirming the inhibiting role played by EO of *F. vulgare*. In addition, the EO emitter did not influence negatively the sensory profile of RTC globe artichoke slices after microwave cooking. Control and active-packaged samples were considered similar by panelists, with the exception of fennel odor and flavor that of course were revealed in samples packed with EO emitter.

Our findings demonstrate the potential application of a *F. vulgare* EO emitter, within an active package system, as a promising technological strategy for preserving quality attributes and reducing microbial spoilage of RTC globe artichoke slices.

Supplementary Materials: The following are available online at https://www.mdpi.com/2304-8158/10/3/517/s1, Figure S1: Experimental work-flow of Ready-to-Cook (RTC) globe artichoke slices.

Author Contributions: Conceptualization, C.R., G.M. (Giuseppe Muratore) and G.M. (Giovanni Mauromicale); methodology, C.R., G.M. (Giuseppe Muratore), and G.M. (Giovanni Mauromicale); investigation, V.R., S.L., R.N.B., G.P. and A.M.; writing—original draft preparation, V.R., C.R., R.N.B. and S.L.; writing—review and editing, V.R., C.R. and S.L; supervision, C.R., G.M. (Giuseppe Muratore) and G.M. (Giovanni Mauromicale); project administration, C.R., G.M. (Giuseppe Muratore) and G.M. (Giovanni Mauromicale); funding acquisition, G.M. (Giovanni Mauromicale). All authors have read and agreed to the published version of the manuscript

Funding: This research was funded by "Utilizzo integrato di approcci tecnologici innovativi per migliorare la shelf-life e preservare le proprietà nutrizionali di prodotti agroalimentari (SHELF-LIFE)", project cod. PON02_00451_3361909.

Data Availability Statement: Main data are contained within the article; further raw data obtained in this study are available on request from the corresponding author.

Conflicts of Interest: The authors declare no conflict of interest.

References

1. Lombardo, S.; Pandino, G.; Mauromicale, G. The influence of pre-harvest factors on the quality of globe artichoke. *Sci. Hortic.* **2018**, *233*, 479–490. [CrossRef]
2. Pandino, G.; Mauromicale, G. Globe artichoke and cardoon forms between traditional and modern uses. *Acta Hortic.* **2020**, *1284*, 1–18. [CrossRef]
3. Pandino, G.; Lombardo, S.; Moglia, A.; Portis, E.; Lanteri, S.; Mauromicale, G. Leaf polyphenol profile and SSR-based fingerprinting of new segregant *Cynara cardunculus* genotypes. *Front. Plant Sci.* **2015**, *5*, 1–7. [CrossRef]
4. Pandino, G.; Lombardo, S.; Mauromicale, G. Chemical and morphological characteristics of new clones and commercial varieties of globe artichoke (*Cynara cardunculus* var. *scolymus*). *Plant Foods Hum. Nutr.* **2011**, *66*, 291–297. [CrossRef] [PubMed]

5. Lombardo, S.; Restuccia, C.; Muratore, G.; Barbagallo, R.N.; Licciardello, F.; Pandino, G.; Scifò, G.O.; Mazzaglia, A.; Ragonese, F.; Mauromicale, G. Effect of nitrogen fertilisation on the overall quality of minimally processed globe artichoke head. *J. Sci. Food Agric.* **2017**, *97*, 650–658. [CrossRef]
6. Muratore, G.; Restuccia, C.; Licciardello, F.; Lombardo, S.; Pandino, G.; Mauromicale, G. Effect of packaging film and antibrowning solution on quality maintenance of minimally processed globe artichoke heads. *Innov. Food Sci. Emerg. Technol.* **2015**, *31*, 97–104. [CrossRef]
7. Barbagallo, R.N.; Chisari, M.; Spagna, G. Enzymatic browning and softening in vegetable crops: Studies and experiences. *Ital. J. Food Sci.* **2009**, *21*, 3–16.
8. Ricceri, J.; Barbagallo, R.N. Role of protease and oxidase activities involved in some technological aspects of the globe artichoke processing and storage. *LWT Food Sci. Technol.* **2016**, *71*, 196–201. [CrossRef]
9. Licciardello, F.; Pandino, G.; Barbagallo, R.N.; Lombardo, S.; Restuccia, C.; Muratore, G.; Mazzaglia, A.; Strano, M.G.; Mauromicale, G. Quality traits of ready-to-use globe artichoke slices as affected by genotype, harvest time and storage time. Part II: Physiological, microbiological and sensory aspects. *LWT Food Sci. Technol.* **2017**, *79*, 554–560. [CrossRef]
10. Pandino, G.; Barbagallo, R.N.; Lombardo, S.; Restuccia, C.; Muratore, G.; Licciardello, F.; Mazzaglia, A.; Ricceri, J.; Pesce, G.R.; Mauromicale, G. Quality traits of ready-to-use globe artichoke slices as affected by genotype, harvest time and storage time. Part I: Biochemical and physical aspects. *LWT Food Sci. Technol.* **2017**, *76*, 181–189. [CrossRef]
11. Rizzo, V.; Lombardo, S.; Pandino, G.; Barbagallo, R.N.; Mazzaglia, A.; Restuccia, C.; Mauromicale, G.; Muratore, G. Shelf-life study of ready-to-cook slices of globe artichoke 'Spinoso sardo': Effects of anti-browning solutions and edible coating enriched with *Foeniculum vulgare* essential oil. *J. Sci. Food Agric.* **2019**, *99*, 5219–5228. [CrossRef]
12. Rizzo, V.; Muratore, G. The Application of Essential Oils in Edible Coating: Case of Study on Two Fresh Cut Products. *Int. J. Clin. Nutr. Diet.* **2020**, *6*, 149. [CrossRef]
13. Wilson, M.D.; Stanley, R.A.; Eyles, A.; Ross, T. Innovative processes and technologies for modified atmosphere packaging of fresh and fresh-cut fruits and vegetables. *Crit. Rev. Food Sci. Nutr.* **2017**, *25*, 1–12. [CrossRef] [PubMed]
14. Dhiman, R.; Aggarwal, N.K. Efficacy of plant antimicrobials as preservative in food. In *Food Preservation and Waste Exploitation*; Socaci, S.A., Frca, A.C., Aussenac, T., Laguerre, J.-C., Eds.; IntechOpen: London, UK, 2019; pp. 1–19. [CrossRef]
15. Scavo, A.; Pandino, G.; Restuccia, C.; Parafati, L.; Cirvilleri, G.; Mauromicale, G. Antimicrobial activity of cultivated cardoon (*Cynara cardunculus* L. var. altilis DC.) leaf extracts against bacterial species of agricultural and food interest. *Ind. Crops. Prod.* **2019**, *129*, 206–211. [CrossRef]
16. Wieczyńska, J.; Cavoski, I. Antimicrobial, antioxidant and sensory features of eugenol, carvacrol and trans-anethole in active packaging for organic ready-to-eat iceberg lettuce. *Food Chem.* **2018**, *259*, 251–260. [CrossRef] [PubMed]
17. La Pergola, A.; Restuccia, C.; Napoli, E.; Bella, S.; Brighina, S.; Russo, A.; Suma, P. Commercial and wild sicilian *Origanum vulgare* essential oils: Chemical composition, antimicrobial activity and repellent effects. *J. Essent. Oil Res.* **2017**, *29*, 451–460. [CrossRef]
18. Restuccia, C.; Conti, G.O.; Zuccarello, P.; Parafati, L.; Cristaldi, A.; Ferrante, M. Efficacy of different citrus essential oils to inhibit the growth and B1 aflatoxin biosynthesis of *Aspergillus flavus*. *Environ. Sci. Pollut. Res. Int.* **2019**, *26*, 31263–31272. [CrossRef]
19. Ruiz-Navajas, Y.; Viuda-Martos, M.; Sendra, E.; Pérez-Álvarez, J.A.; Fernández López, J. Chemical characterization and antibacterial activity of *Thymus moroderi* and *Thymus piperella* essential oils, two Thymus endemic species from southeast of Spain. *Food Control* **2012**, *27*, 294–299. [CrossRef]
20. Diao, W.R.; Hu, Q.P.; Zhang, H.; Xu, J.G. Chemical composition, antibacterial activity and mechanism of action of essential oil from seeds of fennel (*Foeniculum vulgare* Mill.). *Food Control* **2014**, *35*, 109–116. [CrossRef]
21. Salami, M.; Rahimmalek, M.; Ehtemam, M.H.; Szumny, A.; Fabian, S.; Matkowski, A. Essential oil composition, antimicrobial activity and anatomical characteristics of *Foeniculum vulgare* Mill. fruits from different regions of Iran. *J. Essent. Oil Bear. Plants* **2016**, *19*, 1614–1626. [CrossRef]
22. Gulfraz, M.; Mehmood, S.; Minhas, N.; Jabeen, N.; Kausar, R.; Jabeen, K.; Arshad, G. Composition and antimicrobial properties of essential oil of *Foeniculum vulgare*. *Afr. J. Biotechnol.* **2008**, *7*, 4364–4368. Available online: http://www.academicjournals.org/AJB (accessed on 4 January 2021).
23. Chang, S.; Abbaspour, H.; Nafchi, A.M.; Heydari, A.; Mohammadhosseini, M. Iranian *Foeniculum vulgare* essential oil and alcoholic extracts: Chemical composition, antimicrobial, antioxidant and application in olive oil preservation. *J. Essent. Oil Bear. Plants* **2016**, *19*, 1920–1931. [CrossRef]
24. Belabdelli, F.; Piras, A.; Bekhti, N.; Falconieri, D.; Belmokhtar, Z.; Merad, Y. Chemical Composition and Antifungal Activity of *Foeniculum vulgare* Mill. *Chem. Afr.* **2020**, *3*, 323–328. [CrossRef]
25. Soil Survey Staff. *Soil Taxonomy: A Basic System of Soil Classification for Making and Interpreting Soil Surveys*, 2nd ed.; U.S. Gov. Print: Washington, DC, USA, 1999.
26. Pandino, G.; Lombardo, S.; Lo Monaco, A.; Mauromicale, G. Choice of time of harvest influences the polyphenol profile of globe artichoke. *J. Funct. Foods* **2013**, *5*, 1822–1828. [CrossRef]
27. Pandino, G.; Lombardo, S.; Mauro, R.P.; Mauromicale, G. Variation in polyphenol profile and head morphology among clones of globe artichoke selected from a landrace. *Sci. Hortic.* **2012**, *138*, 259–265. [CrossRef]
28. Mauromicale, G.; Ierna, A. Characteristics of heads of seed-grown globe artichoke [*Cynara cardunculus* L. var. *scolymus* (L.) Fiori] as affected by harvest period, sowing date and gibberellic acid. *Agronomie* **2000**, *20*, 197–204. [CrossRef]

29. Rizzo, V.; Amoroso, L.; Licciardello, F.; Mazzaglia, A.; Muratore, G.; Restuccia, C.; Lombardo, S.; Pandino, G.; Strano, M.G.; Mauromicale, G. The effect of sous vide packaging with rosemary essential oil on storage quality of fresh-cut potato. *LWT-Food Sci. Technol.* **2018**, *94*, 111–118. [CrossRef]
30. Licciardello, F.; Lombardo, S.; Rizzo, V.; Pitino, I.; Pandino, G.; Strano, M.G.; Muratore, G.; Restuccia, C.; Mauromicale, G. Integrated agronomical and technological approach for the quality maintenance of ready-to-fry potato sticks during refrigerated storage. *Postharvest Biol. Technol.* **2018**, *136*, 23–30. [CrossRef]
31. Brand-Williams, W.; Cuvelier, M.E.; Berset, C. Use of a free-radical method to evaluate antioxidant activity. *LWT Food Sci. Technol.* **1995**, *28*, 25–30. [CrossRef]
32. Espín, J.C.; Tudela, J.; García-Cánovas, F. Monophenolase activity of polyphenol oxidase from artichoke (*Cynara scolymus*, L.) heads. *LWT Food Sci. Technol.* **1997**, *30*, 819–825. [CrossRef]
33. Espín, J.C.; Tudela, J.; García-Cánovas, F. 4-hydroxyanisole: The most suitable monophenolic substrate for determining spectrophotometrically the monophenolase activity of polyphenol oxidase from fruits and vegetables. *Anal. Biochem.* **1998**, *259*, 118–126. [CrossRef]
34. *UNI EN ISO 13299. Sensory Analysis—Methodology—General Guidance for Establishing a Sensory Profile*; Ente Nazionale Italiano di Unificazione: Milano, Italy, 2016.
35. *UNI EN ISO 8586. Sensory Analysis—General Guidelines for the Selection Training and Monitoring of Selected Assessors*; Ente Nazionale Italiano di Unificazione: Milano, Italy, 2012.
36. *UNI EN ISO 8589. Analisi Sensoriale—Guida Generale per la Progettazione di Locali di Prova*; Ente Nazionale Italiano di Unificazione: Milano, Italy, 2014.
37. Ghidelli, C.; Mateos, M.; Rojas-Argudo, C.; Pérez-Gago, M.B. Novel approaches to control browning of fresh-cut artichoke: Effect of a soy protein-based coating and modified atmosphere packaging. *Postharvest Biol. Technol.* **2015**, *99*, 105–113. [CrossRef]
38. Rico, D.; Martìn-Diana, A.B.; Barat, J.M.; Barry-Ryan, C. Extending and measuring the quality of fresh-cut fruit and vegetables: A review. *Trends Food Sci. Technol.* **2007**, *18*, 373–386. [CrossRef]
39. Lombardo, S.; Restuccia, C.; Pandino, G.; Licciardello, F.; Muratore, G.; Mauromicale, G. Influence of an O_3-atmosphere storage on microbial growth and antioxidant contents of globe artichoke as affected by genotype and harvest time. *Innov. Food Sci. Emerg. Technol.* **2015**, *27*, 121–128. [CrossRef]
40. Pandino, G.; Lombardo, S.; Lo Monaco, A.; Ruta, C.; Mauromicale, G. In vitro micropropagation and mycorrhizal treatment influences the polyphenols content profile of globe artichoke under field conditions. *Food Res. Int.* **2017**, *99*, 385–392. [CrossRef]
41. Restuccia, C.; Lombardo, S.; Pandino, G.; Licciardello, F.; Muratore, G.; Mauromicale, G. An innovative combined water ozonisation/O_3-atmosphere storage for preserving the overall quality of two globe artichoke cultivars. *Innov. Food Sci. Emerg. Technol.* **2014**, *21*, 82–89. [CrossRef]
42. Chang, T.-S. An updated review of tyrosinase inhibitors. *Int. J. Mol. Sci.* **2009**, *10*, 2440–2475. [CrossRef] [PubMed]
43. Scaglione, D.; Reyes-Chin-Wo, S.; Acquadro, A.; Froenicke, L.; Portis, E.; Beitel, C.; Tirone, M.; Mauro, R.; Lo Monaco, A.; Mauromicale, G.; et al. The genome sequence of the outbreeding globe artichoke constructed de novo incorporating a phase-aware low-pass sequencing strategy of F1 progeny. *Sci. Rep.* **2016**, *6*, 19427. [CrossRef]
44. Acquadro, A.; Barchi, L.; Portis, E.; Mangino, G.; Valentino, D.; Mauromicale, G.; Lanteri, S. Genome reconstruction in *Cynara cardunculus* taxa gains access to chromosome-scale DNA variation. *Sci. Rep.* **2017**, *7*, 5617. [CrossRef]
45. Lattanzio, V.; Linsalata, V.; Palmeri, S.; Van Sumere, C.F. The beneficial effect of citric and ascorbic acid on the phenolic browning reaction in stored artichokes (*Cynara scolymus* L.) heads. *Food Chem.* **1989**, *33*, 93–106. [CrossRef]
46. Leoni, O.; Calmieri, S.; Lattanzio, V.; Van Sumere, C.F. Polyphenol oxidase from artichoke (*Cynara scolymus* L.). *Food Chem.* **1990**, *38*, 27–39. [CrossRef]
47. Lattanzio, V.; Cardinali, A.; Di Venere, D.; Linsalata, V.; Palmieri, S. Browning phenomena in stored artichoke (*Cynara scolymus* L.) heads: Enzymic or chemical reactions? *Food Chem.* **1994**, *50*, 1–7. [CrossRef]
48. Rodriguez-Lopez, J.N.; Escribano, J.; Garcia-Canovas, F. A continuous spectrophotometric method for the determination of monophenolase activity of tyrosinase using 3-methyl-2-benzothiazolinone hydrazone. *Anal. Biochem.* **1994**, *216*, 205–212. [CrossRef] [PubMed]
49. CNERNA-CNRS. Produits de la IVe gamme. In *La Qualité Microbiologique des Aliments (Maîtrise et Critères)*, 2nd ed.; Jouve, J., Ed.; Polytechnica: Paris, France, 1996; pp. 73–98.
50. Ministere de l'Economie des Finances et du Budget. *Marché Consommation, Produits Vegetaux prets a L'emploi dits de la "IVemme Gamme": Guide de Bonnes Pratique Hygieniques*; Journal Officiel De La Republique Francaise: Paris, France, 1988; pp. 1621–1629.
51. B.O.E. *Real Decreto 3484/2000, de 29 de Diciembre, por el que se Establecen las Normas de Higiene para la Elaboración, Distribución y Comercio de Comidas Preparadas*; Boletín Oficial del Estado: Madrid, Spain, 2001; p. 11.
52. The Commission of the European Communities. *Commission Regulation (EC) No 1441/2007 Amending Regulation (EC) No 2073/2005 on Microbiological Criteria for Foodstuff*; L 322/12; Official Journal of the European Union: Brussels, Belgium, 2007; pp. 12–29.

Article

Biodegradable Active Packaging as an Alternative to Conventional Packaging: A Case Study with Chicken Fillets

Jawad Sarfraz [1,*], Anlaug Ådland Hansen [1], John-Erik Haugen [1], Trung-Anh Le [2], Jorunn Nilsen [3], Josefine Skaret [1], Tan Phat Huynh [2] and Marit Kvalvåg Pettersen [1,*]

1 Nofima-Norwegian Institute of Food, Fisheries and Aquaculture Research, NO-1431 Ås, Norway; anlaug.hansen@nofima.no (A.Å.H.); john-erik.haugen@nofima.no (J.-E.H.); josefine.skaret@nofima.no (J.S.)
2 Laboratory of Molecular Science and Engineerng, Åbo Akademi University, 20500 Turku, Finland; trung-anh.le@abo.fi (T.-A.L.); tan.huynh@abo.fi (T.P.H.)
3 Norner AS, Asdalstrand 291, NO-3962 Stathelle, Norway; jorunn.nilsen@norner.no
* Correspondence: jawad.sarfraz@nofima.no (J.S.); marit.kvalvag.pettersen@nofima.no (M.K.P.)

Abstract: Innovative active packaging has the potential to maintain the food quality and preserve the food safety for extended period. The aim of this study was to discover the effect of active films based on commercially available polylactic acid blend (PLA_b) and natural active components on the shelf life and organoleptic properties of chicken fillets and to find out; to what extent they can be used as replacement to the traditional packaging materials. In this study, commercially available PLA_b was compounded with citral and cinnamon oil. Active films with 300 µm thickness were then produced on a blown film extruder. The PLA_b-based films were thermoformed into trays. Fresh chicken breast fillets were packed under two different gas compositions, modified atmosphere packaging of 60% CO_2/40% N_2, and 75% O_2/25% CO_2 and stored at 4 °C. The effect of active packaging materials and gas compositions on the drip loss, dry matter content, organoleptic properties, and microbial quality of the chicken fillets were studied over a storage time of 24 days. The presence of active components in the compounded films was confirmed with FTIR, in addition the release of active components in the headspace of the packaging was established with GC/MS. Additionally, gas barrier properties of the packages were studied. No negative impact on the drip loss and dry matter content was observed. The results show that PLA_b-based active packaging can maintain the quality of the chicken fillets and have the potential to replace the traditional packaging materials, such as APET/PE trays.

Keywords: biodegradable; active; natural; essential oil; shelf life; antimicrobial; sensory; poultry

Citation: Sarfraz, J.; Hansen, A.Å.; Haugen, J.-E.; Le, T.-A.; Nilsen, J.; Skaret, J.; Huynh, T.P.; Pettersen, M.K. Biodegradable Active Packaging as an Alternative to Conventional Packaging: A Case Study with Chicken Fillets. *Foods* **2021**, *10*, 1126. https://doi.org/10.3390/foods10051126

Academic Editors: Muratore Giuseppe, Valeria Rizzo and Silvana Cavella

Received: 9 March 2021
Accepted: 13 May 2021
Published: 19 May 2021

1. Introduction

Active packaging based on natural active components has gained much attention during last decade. A packaging is termed "active" when it provides functions beyond the traditional protection and inert barrier to the outside environment [1]. The active packaging consistently interacts with the food and the surrounding environment of the food inside package. In recent years, a lot of research efforts have been put in developing active packaging solutions. These packaging with active components, such as antimicrobials, antioxidants, O_2 scavengers, CO_2 emitters/absorbents, moisture regulators, flavor releasers, and absorbers can delay or stop microbial, enzymatic, and oxidative spoilage [2–4]. Based on the source, the active (antioxidant, antimicrobial) components can be classified in to natural or synthetic. Natural plant-based materials, including essential oils (EO) and their major components, have known antimicrobial and antioxidant properties. EO have been in use for long time as flavoring agents for food [5]. Compounds such as linalool, cinnamaldehyde, carvacrol, thymol, and citral, which are major components of cilantro, cinnamon, oregano, thyme, and lemongrass EO, respectively, have the potential to replace some of the synthetic additives. The antimicrobial activity of the EO can be affected by the pH, fat, and protein content of the product [6]. It has been reported that the EO are more effective against Gram positive bacteria compared to Gram negative bacteria [7]. The EO disturb the structure and permeability of the cell membrane, they interact with the proteins and enzymes and disturb the important cell functions for example, proton motive force, electron flow, energy regulation, or synthesis of structural components [6–8].

Polymeric food packaging materials can be broadly divided into biodegradable and nonbiodegradable. Traditionally nonbiodegradable polymers, such as, polyethylene (PE), polypropylene (PP), polyethylene terephthalate (PET), polyamide (PA), etc., are used in food packaging applications. To reduce the environmental impact from these traditional packaging materials, an alternative approach of developing packaging solutions based on biodegradable materials is being pursued. In recent years, several studies have reported biodegradable and edible coatings with active components such as mint, grape fruit peel extracts, thymol, ferulic acid, caffeic acid, and tyrosol incorporated in—for example—guar gum, gelatin, chitosan, polyvinyl alcohol, or polylactic acid (PLA) matrix with profound antioxidant properties [9–11]. Similarly, the antibacterial properties of various edible coatings and biodegradable packaging with active components based on essential oils for example lemongrass, cinnamon, ziziphora clinopodioides, and/or nanoparticles (e.g., copper oxide, zinc oxide, etc.) incorporated in—for example—alginate, Poly (3-hydroxybutyrate-co-3-hydroxyvalerate), chitosan, sodium alginate-carboxymethyl cellulose, or PLA matrix have been reported [12–16].

PLA is a promising biodegradable and renewable thermoplastic polyester. However, it is hygroscopic in nature, brittle, and less heat tolerant, which makes it less attractive for demanding applications such as food packaging [17]. These properties are greatly improved in the new commercially available PLA blend (PLA_b) materials by the addition of copolyester. Though, PLA films incorporating EO and their major components have been reported in the literature with promising antimicrobial results. These active PLA films were very seldom applied to the real products and when applied, the antimicrobial properties are studied for a short period of time [18–20]. This knowledge gap related to the effect of PLA-based active materials on the real products specially during their entire shelf life, limits the use of these materials in practical applications.

The main objective of this study was to explore the effect of active films based on commercially available PLA_b and natural active components on the shelf life and organoleptic properties of chicken fillets and to find out to what extent they can be used as replacement to the traditional packaging materials. The active films have been developed by incorporating citral and cinnamon oil in different amounts in the polymer matrix.

2. Materials and Methods

2.1. Compounding and Film Blowing

The two additives, cinnamon oil and citral were mixed into Ecovio F2224 using an extruder type W&P26, ZSK MCC with L/D 40 (where L is length and D is diameter) from Coperion GmbH, Germany. No other components were added to the blend. The extrusion temperature was set to 180 °C with a flat profile. The temperature profile on the extruder was increasing from 163 °C in the feeding zone to 183 °C at the die. The extrusion speed was 60 rpm, the output 30 kg/h and N_2 flushing was used. The pure Ecovio F2224 was run through the compounder using the same condition. Compounds of Ecovio F2224 with 4% cinnamon oil as well as 4% citral were made. In addition, a sample with both 2% cinnamon oil and 2% citral was made. The EO (cinnamon and citral) concentrations were selected based on previously reported levels in the literature [21,22] and to limitations in processing parameters as reported by others [22].

From the compounds, 300 μm films were produced on a blown film extruder model E 25, BL180/400 from Collin Lab & Pilot Solutions GmbH, Germany. The temperature profile on the extruder was set from 160 °C to 183 °C in the different zones, and the extrusion speed was 100 rpm. The die gap was 1.5 mm and the blow-up ratio was 1:3.

2.2. FTIR

The Attenuated Total Reflectance Fourier Transform infrared (ATR–FTIR) spectra were recorded using a Thermo Scientific Nicolet iS50 Spectrometer. The instrument is equipped with a diamond crystal and a pressure gauge. The spectra were measured from 700 to 2000 cm^{-1} with a resolution of 4.0 cm^{-1} and averaged from 64 scans.

2.3. Packaging and Storage Conditions

Fresh chicken breast fillets were obtained from a Norwegian producer and the packaging was performed within 48 h after slaughtering by 10 different packaging combinations (PC) as listed in Table 1. Selection of fillets to the packages was randomized to achieve an equal/representative number of total viable count of bacteria on the surface. The chicken fillets were packed using a Multivac R145 thermoformer (Multivac, Wolfertschwenden, Germany). The films were thermoformed to a depth of 3.5 cm at 100 °C for 2 s, after warming for 2.5 s. Bottom webs were sealed to a biaxially oriented polyester top web with an amorphous polyester heat seal (Mylar OL 40, Petroplast, Neuss, Germany). The sealing was done at 150 °C for 2 s. The packages were evacuated to 10 mbar before filling with gas for the modified atmosphere packaging (MAP). MAP was done with two different gas mixtures, 60% CO_2/40% N_2 and 75% O_2/25% CO_2 (AGA, Oslo, Norway). Pre-formed APET/PE trays (K 2187-1U, Faerch plast, Holsebro, Denmark) were used as a positive control. A Multivac T200 tray sealer (Multivac, Wolfertschwenden, Germany) was used for packaging of the APET/PE trays. The APET/PE trays were sealed with the same top web as mentioned above. Two fillets weighing approximately 320 g were packed in the PLA-based trays with volume of 980 mL, and two fillets weighing approximately 250 g were packed in APET/PE trays with volume of 825 mL, resulting in initial gas volume to product volume ratio (g/p ratio) in the range of 2.2 to 2.5. The chicken breast fillets were stored in dark at 4 °C with samples taken after 0, 8, 14, 20, and 24 days of storage.

Table 1. List of ten packaging combinations (PC) consisting of five different packaging materials and two distinct gas compositions: MAP1 (60% CO_2/40% N_2) and MAP2 (75% O_2/25% CO_2).

Packaging Combinations (PC)	Material	Packing Method (MAP)-Gas Composition
1	PLA_b	60% CO_2/40% N_2
2	PLA_b	75% O_2/25% CO_2
3	PLA_b + 4%citral	60% CO_2/40% N_2
4	PLA_b + 4%citral	75% O_2/25% CO_2
5	PLA_b + 4%cinnamon oil	60% CO_2/40% N_2
6	PLA_b + 4%cinnamon oil	75% O_2/25% CO_2
7	PLA_b + 2%citral + 2%cinnamon oil	60% CO_2/40% N_2
8	PLA_b + 2%citral + 2%cinnamon oil	75% O_2/25% CO_2
9	Reference: APET/PE	60% CO_2/40% N_2
10	Reference: APET/PE	75% O_2/25% CO_2

2.4. Head Space Gas Chromatography/Mass Spectrometry (GC/MS)

Gas samples were taken by using a syringe to draw 120 mL of the headspace in the packages through an adsorbent tube packed with Tenax GR for trapping the volatile compounds. The adsorbent tubes were desorbed at 280 °C for 7 min in a Markes Thermal Desorber and transferred to an Agilent 6890 GC with an Agilent 5973 Mass Selective Detector (EI, 70 eV). The volatiles were separated on a DB-WAXetr column (30 m, 0.25 mm i.d., 0.5 μm film) with a temperature program starting at 30 °C for 10 min, increasing 1/min to 40 °C, 3/min to 70 °C, and 6.5/min to 230 °C, hold time 5 min. The peaks were integrated, and compounds tentatively identified with HP Chemstation software and NIST 2011 Mass Spectral Library. The volatile compounds are expressed in arbitrary units of the deconvoluted component of the peak area. The result shown are an average of three replicates.

2.5. Gas Transmission Rates

The oxygen transmission rate (OTR) for the different packages was found by the ambient oxygen ingress rate method (AOIR) [23] and the carbon dioxide transmission rate (CO_2TR) was determined as described by Larsen and Liland (2013) [24]. The gas transmission rates were reported as ml gas/package/day. The OTR and CO_2TR were measured at 4 °C, approximately 40% RH. The result shown are an average of four replicates.

2.6. Drip Loss

Drip loss was determined by initially weighing empty packages (A), packages with the fillets at the time of packaging (B), and at each sampling time weighing of (C) the packages including the drip loss liquid after removing the chicken. The amount of drip loss was found by calculating the increase in weight of the packages (C-A) and divided by the initial weight of the chicken: (C-A)/(B-A). Results are given as the percentage (%) of initial muscle weight and refer to the corresponding liquid loss from the meat sample. The result shown are an average of three replicates.

2.7. Dry Matter

Dry matter in chicken breast fillets was determined at each sampling time. The water content was determined in difference of weight before and after oven-drying at 105 °C for 16–18 h. The chicken fillets were homogenized and approximately 6 ± 0.5 g was weighed and stored in petri dishes (diameter 90 mm) before drying. After drying, the petri dishes were cooled to room temperature in an exicator before weighing. The dry matter was calculated as 100% minus water content. The result is given as average of three replicates.

2.8. Microbiology

Selected microbiological analyses were performed at the time of packaging (five replicates) and at each sampling time; after 8, 14, 20, and 24 days of storage (three replicates) as explained earlier [25]. Briefly, from the surface of the fillet, samples of 3 × 3 cm^2 and 1 cm depth were cut with a sterile scalpel, weighed, and added approximately 90 mL peptone water to a 1/10 dilution and run in a stomacher for 60 s.

Appropriate 10-folds dilutions were spread by using Whitley automatic spiral plater (WASP) (Don Whitley Scientific Ltd. Bingley, England) in duplicate on:

1. Plate count agar (PCA; Difco, Difco Laboratories, Detroit, MI, USA) for total viable counts (TVC): incubation temperature 30 °C, 72 h, aerobic incubation. (NMKL No. 86);
2. De Man, Sharpe, and Rogosa agar (MRS; Oxoid, Unipath Ltd., Basingstoke, Hampshire, UK,) for lactic acid bacteria (LAB): 25–30 °C, 48 h, aerobic incubation. (NMKL No. 140);
3. Streptomycin thallous acetate actidione (STAA) agar base (CM 0881 with selective supplement SR 0151E, Oxoid, Hampshire, England) for *Brochothrix thermosphacta*; 25 °C for 48 h, aerobic incubation. (NMKL No. 141);
4. Violet Red Bile Glucose Agar (VRBGA, Oxoid, Hampshire, UK) for *Enterobacteriace*; 37 °C, 24–48 h, semi-aerobic conditions, cells embedded in agar with sterile overlay. (NMKL No. 144).

Microbial counts are expressed as colony forming units (cfu) per g.

2.9. Sensory Analysis

To describe the objective perception of the various samples, a trained panel consisting of 10 assessors, performed a Quality Descriptive Analysis, ISO 13299:2016 (E) of the samples after 14 days of storage. The panelists have been trained according to recommendations in ISO 8586:2012 (E). The sensory laboratory has been designed according to guidelines in

ISO 8589:2007 (E) with separate booths and electronic registration of data (Eye Question, v. 3.8.6, Logic 8, Elst, The Netherlands), standardized light, and a separate ventilation system.

The samples were taken from the cold storage room for tempering two hours prior to evaluation and cut up, alongside, and put into bowls 30 min before evaluation. The samples were 18 ± 2 °C when serving. Raw chicken, one half of a fillet was first served to the assessors in transparent plastic cups covered with lid for evaluation of the odor. Then, the samples were taken out of the cups and placed on a white cardboard plate, and the appearance of the samples was evaluated. Finally, the assessors evaluated texture by pressing the sample on the thickest part with two fingers (using plastic gloves). The coded samples were served in blind trials randomized according to sample, assessor, and replicate.

The panelists evaluated the samples in duplicate, during five sessions with at least fifteen minutes break between each session. The assessors recorded their results at individual speed on a 15 cm non-structured continuous scale with the left side of the scale corresponding to the lowest intensity and the right side corresponding to the highest intensity. The computer transformed the responses into numbers between 1 = low intensity and 9 = high intensity.

2.10. Statistical Analyses

To compare the results obtained from 10 different PC as listed in Table 1, one-way analysis of variance (ANOVA) and comparisons Tukey test (which compares all pairs of groups, while controlling the simultaneous confidence level) was performed at each sampling time. General Linear Model (GLM) was applied with the variables: materials, gas composition, and storage time. Their second order interactions were fitted to the measured responses. Analysis of variance has been conducted in Minitab 18.

The multivariate statistics software package, The Unscrambler® v9.8 (Camom Software AS, Oslo, Norway) was used for the principal component analysis (PCA) and partial least squares regression (PLSR) analysis of the combined chemical and sensory data. Full cross validation (leaving one sample out) and raw and auto scaled preprocessing (variables divide by their standard deviation) of the data were used in the data analysis.

3. Results and Discussion

3.1. FTIR

FTIR was performed in order to confirm the presence of active components in the compounded films. Figure 1 show IR spectra of the PLA_b film and PLA_b-based active films. In the PLA_b-citral sample, the band at 1673 cm^{-1} is particularly noticeable (Figure 1b) due to contribution of the -C=O- stretching vibration of citral [26,27]. On the other hand, the 1514 cm^{-1} band is distinct for the PLA_b-cinnamon oil (Figure 1c) and is believed to belong to -C=C- stretching vibration of the aromatic ring or N-O stretching (of nitro compound left in cinnamon oil) [28]. The band is quite distinct as it is separated from the band of 1504 cm^{-1} which is for the pure PLA_b sample. The IR bands of both citral and cinnamon can also be discriminated for PLA_b film containing 2% citral and 2% cinnamon oil (Figure 1d). Interestingly, their intensity is half of 4% citral (for the 1673 cm^{-1} band) and 4% cinnamon (for the 1514 cm^{-1} band).

These results manifest FTIR as a useful tool for both qualitative and semi-quantitative evaluation of the incorporated active components. The success in qualitative measurement is due to the major difference in the chemical formula (with and without the aromatic ring) of citral and cinnamon. Meanwhile, the concentrations of citral and cinnamon incorporated in the PLA_b film closely correlate with intensity of the IR bands, and therefore allow to be semi-quantified by FTIR.

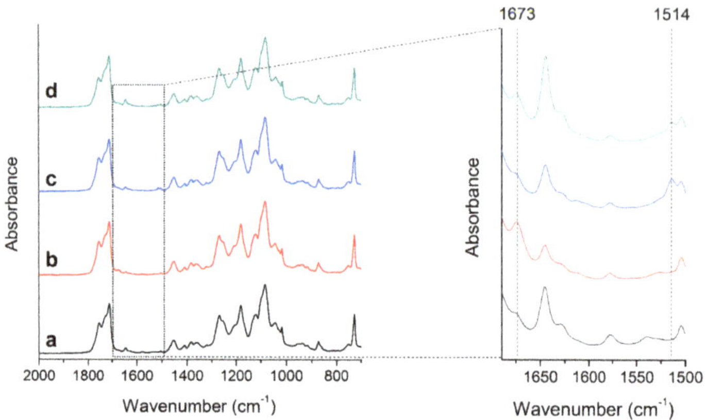

Figure 1. IR spectra of the 300 μm thick PLA_b (**a**) before doping and after incorporating (**b**) 4% citral, (**c**) 4% cinnamon oil, and (**d**) 2% citral and 2% cinnamon oil. The right figure is a zoom of all spectra at wavenumber of 1500–1680 cm^{-1} where significant differences can be seen at 1514 and 1673 cm^{-1}.

3.2. Gas Barrier Properties

Typically, for food packaging applications carbon dioxide transmission rate (CO_2TR) and oxygen transmission rate (OTR) is measured to quantify the gas barrier properties of the polymer.

The Figure 2A,B shows the OTR and CO_2TR of PLA_b-based packages and the reference consisting of APET/PE. The OTR and CO_2TR of the reference APET/PE package was 1.6 ± 0.9 and 3 ± 1.4 mL/package/day, respectively, while for PLA_b package the OTR was 14.4 ± 1.3 and CO_2TR was 51.4 ± 4.6 mL/package/day. Zeid et al. reported OTR for 112 μm PLA to be 865.9 ± 38.9 mL/m^2/day atm [29]. However, OTR is dependent on several aspects such as thermoforming and material thickness, in addition to additives.

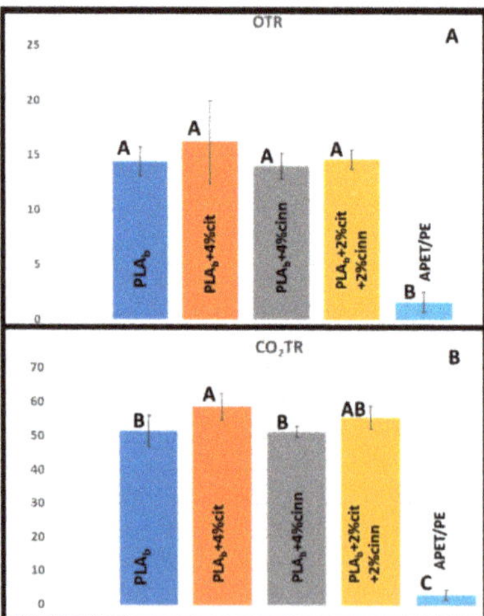

Figure 2. The oxygen transmission rate (**A**) and carbon dioxide transmission rate (**B**) of the PLA_b, PLA_b composites and APET/PE packages. (cit = citral, cinn = cinnamon oil) in ml gas/package/day. Upper case letters represent one-way analysis of variance Tukey comparison, samples which do not share the same letter are significantly different.

Figure 2A,B shows that the PLA_b based packages has relatively poor barrier properties compared to the APET/PE as expected. According to one-way ANOVA, there was no significant difference in the OTR of the PLA_b based packages, which implies that the oxygen barrier properties of PLA_b were not significantly affected by the incorporation of the active components (citral and/or cinnamon oil) corresponding to other findings [28]. However, APET/PE package has, as expected, lower OTR compared to the PLA_b based packages (Figure 2A). On the other hand, the addition of citral in the PLA_b matrix has slightly increased the CO_2TR. (Figure 2B). In contrast, the addition of cinnamon oil did not significantly change the CO_2TR. The differences in OTR for the PLA_b and APET/PE packages in our study were approximately 10-fold. In the literature 100× higher oxygen permeability for PLA has been reported compared to PET. [30] For PLA, 216 cm^3.mm/m^2.day.atm and for PET, 1.2–2.4 cm^3.mm/m^2.day.atm were reported at 23 °C [21]. Similar OTR trends were also reported for PLA and APET/PE [31,32]. However, as described by Pettersen et al., OTR measured for PLA based pouch with 40 µm thickness was 4.75 mL O_2/package/day as compared to 0.031 mL O_2/package/day for thermoformed trays made of 550 µm APET/PE measured at 6 °C [31].

Typically, APET/PE material is used for MAP of chicken breast fillets and when packed under 60% CO_2/40% N_2 (commonly applied gas composition in Norway), the shelf life of approximately 3 weeks can be achieved for chicken fillets [33].

The oxygen content in the PLA_b based packages with chicken stored in 60% CO_2/40% N_2 increased to almost 2% after 14 days of storage, while the level was only 0.1% in the APET/PE reference package during the same time. As expected, the CO_2 concentration in all packages decreased during storage due to solubility of CO_2 into the meat. In PLA_b based packages the level of CO_2 decreased from 60% to approximately 30% after 7 days of storage, with further reduction to around 20% at the end of storage time and for APET/PE the level of CO_2 decreases to 45% after 7 days of storage and stayed almost at the same level to the end of storage time.

In packages with chicken store in 75% O_2/25% CO_2 atmosphere the oxygen content slightly increased in all packages (up to 80% after 14 days of storage in the PLA_b-based packages), followed by a reduction to 70–75%. The level of CO_2 was reduced from initial 22% to approximately 14% after 14 days of storage and an increase to 17–19% after 24 days of storage. For APET/PE the content CO_2 decreased somewhat (from 22 to 20%) after 7 days and increased to 25% at the end of storage time.

The presence of oxygen in the package can have many adverse effects on the product including oxidative spoilage, off odor, and discoloration which limits the shelf life of the product. On the other hand, the use of CO_2 in MAP can suppress the growth of common spoilage bacteria [34], thus improving the shelf life and food safety of the product.

3.3. Drip Loss and Dry Matter

Chicken fillets are prone to drip loss and several factors, such as time of deboning, refrigerator time, and packaging environment, may affect the liquid loss [35]. In order to maintain an attractive product appearance as well as to avoid reduction of sensory quality, such as juiciness, drip loss from the product should be minimized. Earlier findings have shown that MAP with relatively high CO_2 concentration (≥40%) increases the drip loss. This increase in drip loss has been explained by the reduction in pH, resulting in the decrease in water holding capacity of proteins [36,37]. However, Pettersen et al., reported no significant differences in drip loss for chicken stored in high CO_2 concentration (60% CO_2/40% N_2) compared to lower concentration (25% CO_2/75% O_2) [38]. Holck et al., have also shown that the drip loss is not only related to the content of CO_2 in the head space, as high content of CO_2 with addition of a CO_2 emitter resulted in lower drip loss compared to samples with lower content of CO_2 [25,38]. The absorption of CO_2 at the surface of the product may result in under pressure in the package, which may increase the drip loss. However, it also depends on the packaging design and degree of filling (gas/product ratio).

Furthermore, it has been reported in the literature that increased drip loss will take place in rigid packages compared to flexible ones [39].

Chicken fillets were stored by applying ten different packaging combinations (1–10), as listed in Table 1. The drip loss from the product has been measured and the results are shown in Figure 3. According to GLM (with variables: material, gas composition, and storage), neither the material nor the gas composition has any significant effect on the drip loss in our study (Table S1). On the other hand, storage time as well as the interaction between material and gas composition has a significant effect on the drip loss as explained by 40.5% and 7.1% of the variance, respectively. Figure 3 shows that the drip loss increased slightly with the storage time. Nevertheless, a maximum drip loss of around 2% was measured at the end of 24 days. These results are in agreement with the previously reported values in the literature under similar conditions. Holck et al. has reported a drip loss of around 2.5% for chicken fillets stored under MAP (60% CO_2/40% N_2) at the end of 26 days. According to one-way Anova performed for each sampling time, only after 8 days of storage, significant differences between some of the packaging combinations of PLA_b were observed, as shown in Figure 3. Furthermore, storage of chicken in APET/PE with 60% CO_2 (PC 9) resulted in higher drip loss compared to PC 5 (PLA_b + 4% cinnamon oil with 60% CO_2) and were placed in different Tukey groups as shown in Figure 3. For all other sampling time (14, 20, and 24 days) no significant differences between the samples were observed. Figure 3 shows that the PLA_b-based active packaging did not contribute to any increase in the drip loss which is further confirmed by the GLM analysis. Pure PLA is hygroscopic in nature [17], which could be of importance when studying the interaction of this material with high water content products. In this regard, drip loss, appearance, and surface dryness are important parameters to study [40].

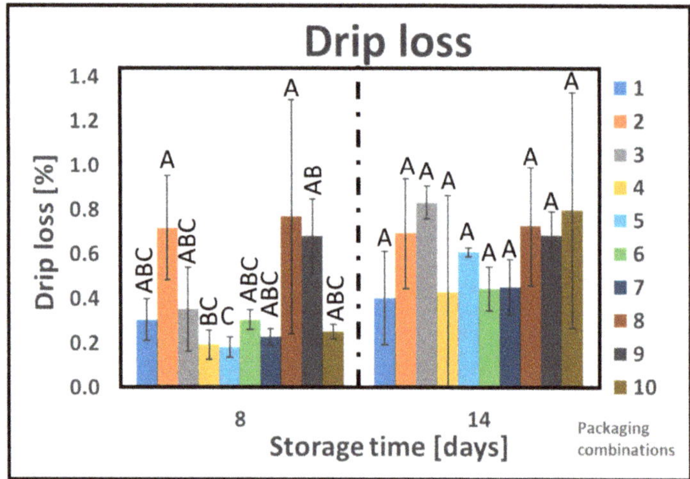

Figure 3. Drip loss of chicken samples as a function of storage time for different packaging combinations. One-way analysis of variance Tukey comparison is performed separately at different sampling points. Tukey group is represented by upper case letters, samples which do not share the same letter are significantly different.

The moisture and dry matter content are important guides for the meat quality. The dry matter content is positively correlated with total nutrient content [41]. Shaarani et al. (2006) reported moisture content in broiler meat being as high as 76% [42]. The dry matter content results are shown in Figure S1. According to statistical analysis, GLM, material, and storage time has a significant effect on the dry matter content and explained variance of 8.8% and 12.3%, respectively. However, residual variation was relatively high, 65% and residual R^2 (adj) only 15.5%, meaning that much of the variance was not explained by the model. It can be seen (Figure S1) that at any given sampling time, the variation in dry matter content between the samples is less than 1%, this difference is probably not of practical importance. According to one-way Anova performed for each sampling time (8, 14, 20, and 24 days), no significant differences between the samples were observed. Furthermore, the dry matter content of 24.5 ± 1% independent of packaging combination is in agreement with the reported values in the literature [42].

3.4. Microbiological Growth

The initial level of bacteria in the fresh chicken breast fillets measured as TVC was about 2.8 log cfu/g at the time of packaging.

Figure 4 shows the effect of the ten different packaging combinations on the microbial quality measured as TVC of the chicken fillets. According to the GLM (Table S1), storage time has significant effect on the microbial growth, as expected. TVC in all samples increased from the initial level of about 2.8 log cfu/g to above 6 log cfu/g during the first 14 days of storage (Figure 4). According to the literature, spoilage can be found at total bacterial count levels of above 7 log cfu/g, which indicates that samples stored in some of the packaging combinations presented had passed its microbiological shelf life already after 14 days of storage [37,43,44]. During the first 14 days of storage, under 60% CO_2/40% N_2 all the chicken samples packed in PLA_b and PLA_b composites have higher growth of bacteria compared to the reference APET/PE. However, opposite trend was observed under 75% O_2/25% CO_2.

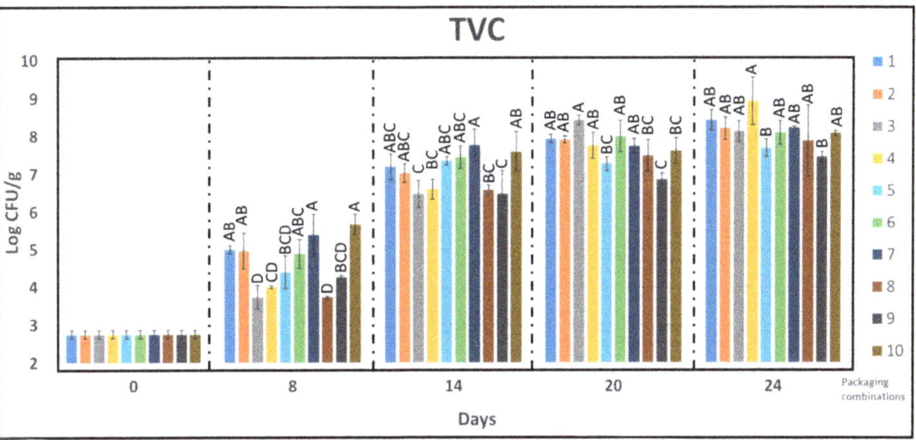

Figure 4. The effect of packaging combinations on the total viable counts (log cfu/g) in chicken samples during storage time of 24 days. One-way analysis of variance Tukey comparison is performed separately at sampling point 8, 14, 20, and 24. Tukey group is represented by upper case letters, samples which do not share the same letter are significantly different.

According to the Tukey pair wise comparison of the TVC at the end of 8 days, lowest level of TVC (3.7 log cfu/g) was obtained for some of the packaging combinations containing essential oils (Figure 4). Furthermore, the TVC results show that the microbiological quality of the chicken meat packed in APET/PE package was no different than those packed in PLA_b-based active packages despite relatively poor gas barrier properties of these materials.

It can be concluded from the TVC results that, during the first 8 days of storage, some effect of active compounds is obtained as inhibition in the growth of microorganisms has been achieved from PC 3, 4, and 8 (all PLA_b with added citral) compared to the positive control (PC 9; APET/PE with 60%CO_2 / 40%N_2). It can correspond to the sustained release of active components (further confirmed by GC/MS Section 3.5). PC 3 (PLA_b added 4%

citral) has significant lower level of TVC compared to neat PLA_b (PC 1) and PC 7 (PLA_b added 2%Citral +2% cinnamon) after 8 and 14 days of storage. Furthermore, the TVC results from sampling time 14, 20, and 24 days confirmed that PLA_b-based packaging (PC 3, 4, 5, and 8) are similar compared to the positive control with APET/PE and 60% CO_2/40% N_2 (PC 9) in maintaining the microbiological quality of the chicken at different time points. The reduction of antimicrobial activity as a function of storage time for natural antimicrobials has been reported in the literature [45]. As the antimicrobial activity of the essential oils is based on the release of volatile active components. With the passage of time, due to decrease in the concentration of the active components, their antimicrobial activity decreases (further refer to Section 3.5).

The growth of three common spoilage bacteria, Lactic acid bacteria (LAB), *Brochothrix thermosphacta*, and *Enterobacteriaceae* were further studied in detail for the chicken samples. Figure 5 shows the effect of different packaging combinations on the growth of LAB. During storage, lactic acid bacteria increased for all samples. This is in accordance with other studies of modified atmosphere packaged poultry products [37,46,47].

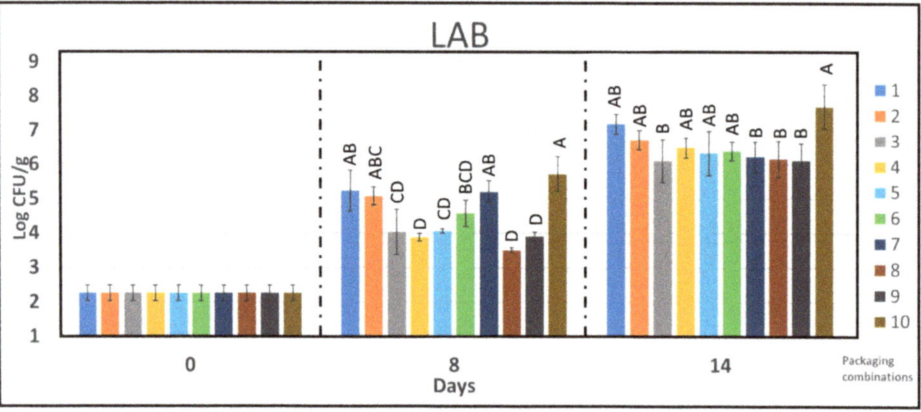

Figure 5. The effect of packaging combinations on the LAB counts in chicken samples during storage time of 14 days. One-way analysis of variance Tukey comparison is performed separately at sampling point 8 and 14 days. Tukey group is represented by upper case letters, samples which do not share the same letter are significantly different.

Under 60% CO_2/40% N_2, the chicken fillets stored in active PLA_b composites and APET/PE has similar bacterial load (around 10^6 cfu/g) after 14 days of storage. However, the LAB count from reference PLA_b sample was an order of magnitude higher under the same conditions. Under MAP2 (75% O_2/25% CO_2), all PLA_b samples have lower LAB counts compared to APET/PE during the 14 days of storage.

After 8 days storage of chicken, PC 4 (PLA_b with citral) and PC 8 (PLA_b with citral+cinnamon) were similar in suppressing the growth of LAB as compared to control PC 9 and belong to the same Tukey group, as shown in Figure 5. Overall, PLA_b containing citral has shown some inhibitory effect towards the growth of lactic acid bacteria after 8 days of storage. However, during further storage the effect diminishes (Figure 5).

The count numbers of *Brochothrix thermosphacta* were between 2 to 4 log cfu/g after 14 days of storage, as shown in Figure 6. However, both APET/PE reference and PLA_b-based packages suppressed the growth of *Brochotrix thermosphacta* under 60% CO_2/40% N_2 (Figure 6). Figure 6 further shows that PC 3, 5, and 9 extended the lag phase of *Brochotrix thermosphacta* for up to 14 days under 60% CO_2/40% N_2 and during this time the concentration of bacteria remained similar for these materials. It is reported in the literature that increasing concentrations of CO_2 retards the growth of *Brochothrix thermosphacta* [37].

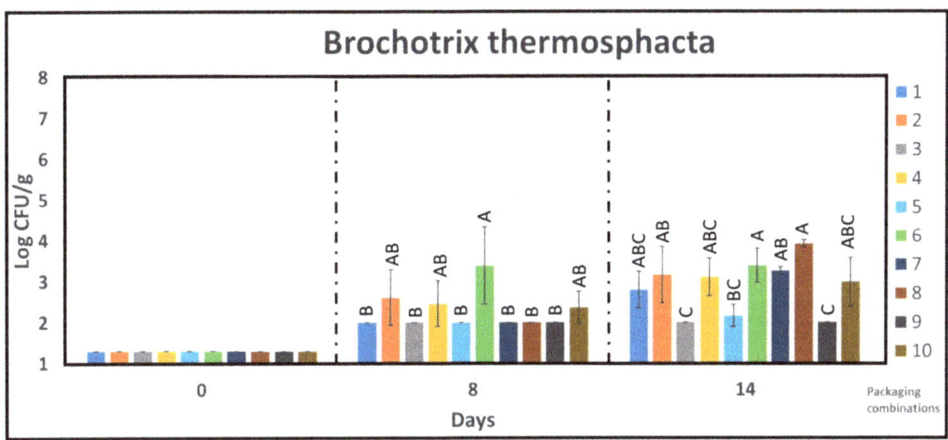

Figure 6. The effect of packaging combinations on the *Brochotrix thermosphacta* counts in chicken samples during storage time of 14 days. One-way analysis of variance Tukey comparison is performed separately at sampling point 8 and 14. Tukey group is represented by upper case letters, samples which do not share the same letter are significantly different.

At 8 days of storage, no significant growth of *Brochothrix thermosphacta* was observed, as shown in Figure 6, and according to one-way Anova, no significant difference between the chicken samples from different PC was obtained except for PC 6 as shown in Figure 6. At 14 days, PLA_b in 60% CO_2/40% N_2 (PC 3, 5, and 9) had the lowest mean count of *Brochothrix thermosphacta* (same Tukey group). Although significant difference between the samples were observed after 14 days of storage (4 different Tukey group), the mean count was less than 4 log cfu/g for all the samples.

Generally, PLAb and PLAb-based active packaging with high O_2 was more effective in suppressing the growth of Enterobacteriaceae, as shown in Figure 7. During the first 14 days of storage, under 60% CO_2/40% N_2 all the chicken samples packed in PLAb and active PLAb composites had higher growth of bacteria compared to the APET/PE (PC 9). However, opposite trend was observed under 75% O_2/25% CO_2. The distribution of samples according to different Tukey group at different storage time for Enterobacteriaceae is also shown in Figure 7.

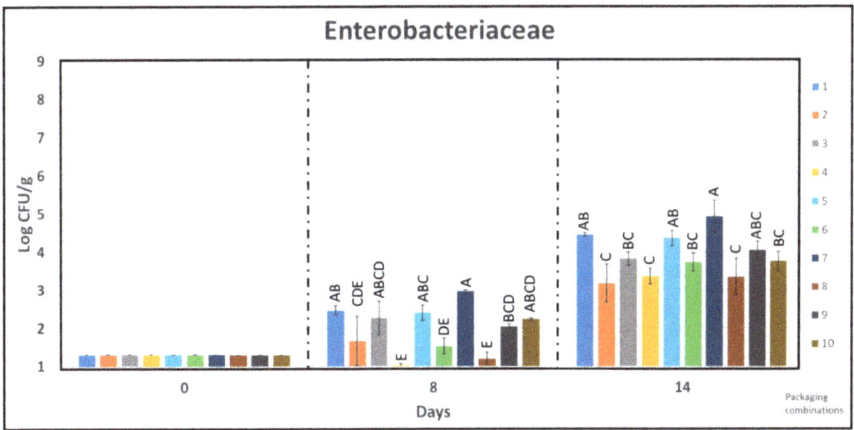

Figure 7. The effect of packaging combinations on the *Enterobacteriaceae* counts in chicken samples during storage time of 14 days. One-way analysis of variance Tukey comparison is performed separately at sampling point 8 and 14. Tukey group is represented by upper case letters, samples which do not share the same letter are significantly different.

3.5. GC/MS

The GC/MS analysis of the empty packaging shows that volatile compound associated with solvent i.e., tetrahydrofuran was dominated in the PC 1, 3, 5, and 7, as shown in Figure 8. Identified main volatile compounds in the empty packaging and their relative concentrations are compiled in Table S3. PLA_b incorporating cinnamon oil (PC 5 and 7) was dominated by the typical major terpenes found in cinnamon oil, i.e., α-pinene, α-phellandrene, o-cymene, and camphene and low levels of the other typical cinnamon oil terpenes [48]. The PLA_b containing citral (PC 3 and 7) was dominated by the two major isomers of citral, tr(α)-citral and cis(β)-citral, and low levels of other citral derived terpenes. The levels of these major volatiles in the active packaging materials showed a linear decrease in the range of 20–60% with storage time.

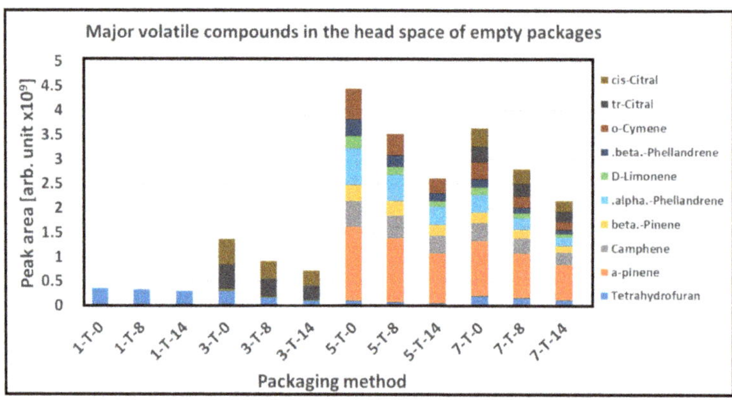

Figure 8. Major volatile compounds in the head space of empty packages with PC 1, 3, 5, and 7 with storage time of 0, 8, and 14 days. (x-T-y; x = packaging method, y = storage time in days).

Similarly, GC/MS head space analysis of the packaging with chicken samples also shows the strong presence of major terpenes found in cinnamon oil, i.e., α-pinene, β-pinene, α-phellandrene, β-phellandrene, o-cymene, D-limonene, and camphene as well as two major isomers of citral, tr(α)-citral, and cis(β)-citral, as shown in Figure 9. The major cinnamon oil terpenes showed a decrease over 14 days storage in the PLA_b + cinnamon oil samples (PC 5 and 6), but in the combined cinnamon oil and citral packaging material (PC 7 and 8), this effect was not that pronounced as shown in Figure 9. The PLA_b + citral sample showed a slight increase (PC 3), while the PC 4 showed a slight decrease with storage time in the citral isomers from 8 to 14 days, whereas the combined PLA_b cinnamon oil and citral samples (PC 7 and 8), showed a decrease in the citral isomers (Figure 9).

Figure 10 shows the major volatile bacterial metabolites after the storage time of 14 days. Bacterial secondary metabolites dominated by dimethyl sulphide, acetone, dimethyl sulfoxide, carbon sulphide, acetic acid, and ethanol were detected, as shown in Figure 10. No typical secondary lipid oxidation products from chicken could be found, which could otherwise be expected in the MAP2. These results are in good agreement with the literature. M. Eilamo et al. detected butene, ethanol, acetone, pentane, dimethylsulfide, carbon disulfide, and dimethyl disulfide as primary volatile compounds in gas packed poultry meat with gas composition of oxygen 0–4%, carbon dioxide 20, 50, and 80% with nitrogen being the balancing gas [49].

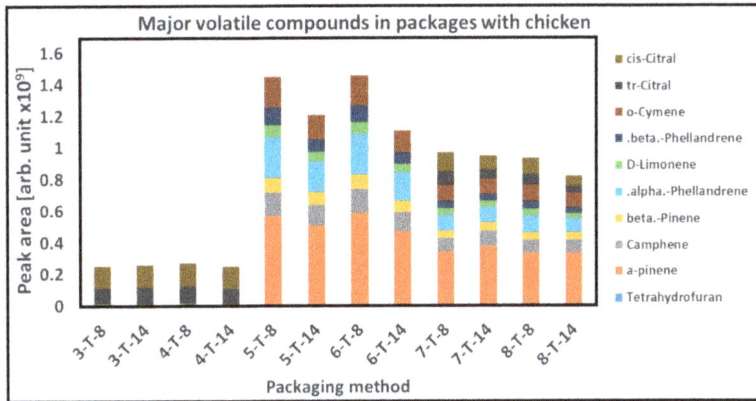

Figure 9. Major volatile compounds in the packaging headspace containing chicken fillets packed with different packaging combinations (PC 3, 4, 5, 6, 7, and 8) after 8 and 14 days of storage. (x-T-y; x = packaging method, y = storage time in days).

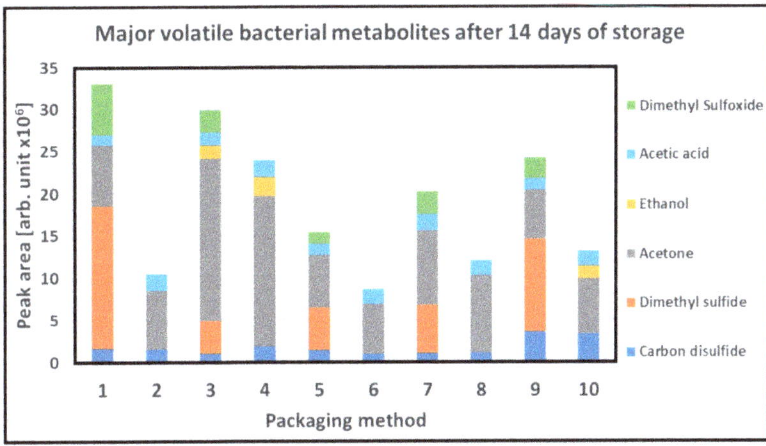

Figure 10. Major volatile bacterial metabolites in the packaging headspace containing chicken fillets packed with different packaging combinations after 14 days of storage.

Upon microbial metabolism, generally volatile organic compounds are produced via two biochemical pathways: glycolytic and proteolytic. Ethanol can be produced from carbohydrates via anaerobic fermentation, or via deamination and decarboxylation of amino acids. In meat, the ethanol production is often attributed to the metabolic activity of lactic acid bacteria. [50–52] Sulfides are produced through proteolytic pathway and their production is ascribed to the metabolic activity of *Pseudomonas*, *Enterobacter*, Acinetobacter/Moraxella, and Alteromonas putrefaciens [53].

Furthermore, our results indicate a difference in spoilage between the packaging gases with high levels of dimethyl sulphide and dimethyl sulfoxide for the MAP1 variants compared to the MAP2 variants. This also corroborates with the literature. Dimethyl sulfide, dimethyl disulfide, dimethyl trisulfide, and carbon disulfide has been reported as volatile biomarkers in chicken hindquarters packed under a modified atmosphere [49,54] however under aerobic conditions they are not suitable for early spoilage detection [51]. It can be further seen in Figure 10 that sample from active-PLA$_b$ based packaging has significantly lower levels of dimethyl sulfide. The higher levels of acetone found in the citral containing PLA_b, may partly be ascribed to the oxidation of citral, since acetone is an oxidation product of citral.

Detected major volatile compounds in the packaging with chicken samples with their relative concentrations are listed in Table S4.

3.6. Sensory Profile

The chicken samples stored for 14 days were analyzed for odor, appearance, and texture by a trained sensory panel with ten professional assessors using a quantitative descriptive method, ISO 13299:2016E. Definition of sensory attributes is provided in Table 2. ANOVA of the sensory evaluation data revealed that there were significant differences between the chicken variants for all the odor attributes, except for metallic, rancid, and prickly as shown in Figure 11. Odor intensity below 3 is considered low and is generally believed to be not recognizable by majority of the consumers. For appearance and texture attributes, there were statistically significant differences for color hue, color intensity, and gloss. (Table S2) In general, there were greater variations between the samples for the odor attributes than for the appearance and the texture attributes. In addition, surface degradation or surface slime formation was not seen on any of the samples stored with different PC when analyzed at the end of 14 days. Along with oxidative spoilage, the generation and accumulation of metabolic by-products upon enzymatic degradation of food can also lead to discoloration, texture change and development of off-odor. Offensive odor is developed in meat when bacterial flora at the surface exceeds 10^7 cfu/cm^2 and above10^8 cfu/cm^2 the surface becomes slimy [37,43,44,55]. Interestingly no differences were recorded for dried appearance, dried texture, and firmness. PLA is hygroscopic and is expected to have a drying effect on the product with high water content, such as poultry [17,32]. However, ecovio® F2224 used in this study is a compound of biodegradable copolyester ecoflex® F Blend and polylactic acid (PLA, NatureWorks®), which explains the improved properties of the material.

Principal component analysis (PCA) showed that PCA1 and PCA2 described 47 and 23% of the variation among the samples, respectively (Figure S2). Lemon and sour odor versus fermented, cloying, and sulfur odor described most of the variations along PCA1 and PCA2. The Biplot further shows similarities in the negative odor attributes between the samples stored in PC 1, 2, 9, and 10 (reference PLA_b and APET/PE) with good correlation to cloying, fermented, and sulfur odor (Figure S2). On the opposite direction the PLA_b variants with additives correlated to sour odor (PC 3, 4, 7, 8), lemon (PC 3, 4, 7, 8), and cinnamon odor (PC 5, 6).

Table 2. Definition of sensory attributes used in sensory profiling of raw chicken.

Odor Attribute	Description
Sour odor	Related to a fresh, balanced odor due to the presence of organic acids
Burnt odor	Associated with a burnt/burning odor
Metallic odor	Odor of metal (ferrous sulfate)
Sulfur odor	Odor of sulfur
Cinnamon odor	Associated to odor of spices (cinnamon, cloves, nutmeg)
Lemon odor	Associated to odor of lemon/ lemon aroma
Off odor	Related to odor which does not naturally exist in chicken
Cloying odor	Associated with an unfresh, sickening odor
Fermented odor	Associated to fermented acids, tainted
Rancid odor	The intensity of all rancid odors (grass, hay, candle, paint)
Prickly odor	Associated with a sharp, pungent odor

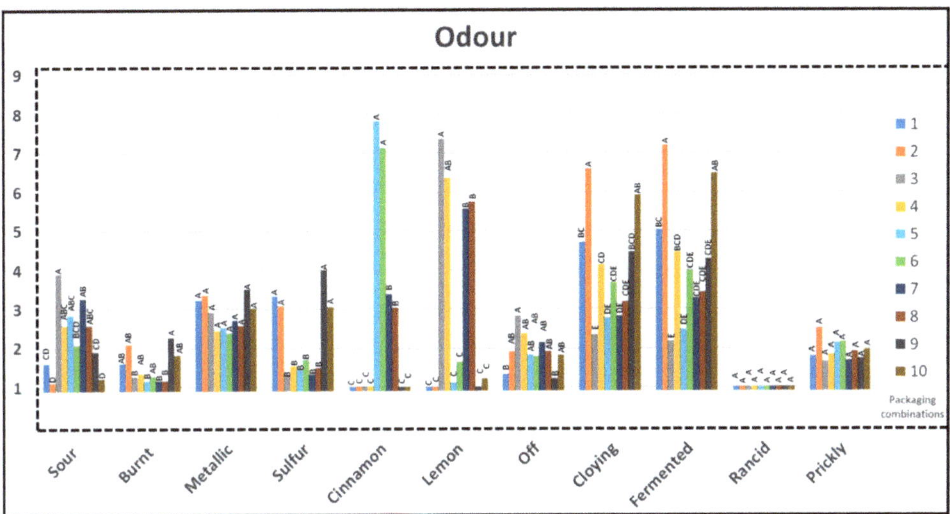

Figure 11. Mean intensities of the assessed odor attributes of all samples after 14 days of storage. Tukey group is represented by upper case letters, samples which do not share the same letter are significantly different.

Sour odor is usually experienced positively by the consumer and is related to a fresh chicken with good quality. On the contrary, fermented, cloying, burnt, sulfur, metallic, off-odor, rancid, and prickly odors are experienced as negative attributes and are related to spoiled chicken with a poor quality. Negative odor attributes cloying and fermented received high scores for samples packed in PC 1, 2, 4, 6, 9, and 10. The mean value for sulfur odor was more than 3 for chicken samples from PC 1 and 9 which can be related to the amount of detected dimethyl sulfide (Figure 10). It can be noted that although dimethyl sulfide was detected for samples from PC 3, 5, and 7 (Figure 10), mean sensory score for sulfur was however much lower for these samples compared to samples stored in PC 2 and 10. It can be implied that the additive (citral and cinnamon oil) might have partially masked the negative sulfur odor in this case.

3.7. Volatiles/Sensory

The results from a PLS regression model based on the volatile compounds (x-data) and sensory data (y-data) showed the following relationship between the volatile compounds and the respective sensory odor attributes: As would be expected, the odor attribute sulfur is associated with the volatile sulfur compounds (dimethyl sulfide, carbon disulfide and dimethyl sulfoxide) and to a lesser extent, also the attributes burnt and metallic are associated with these substances, which may be due to the fact that these two attributes may have burnt and metallic notes. Cinnamon odor is strongly associated with tr-cinnamaldehyde, in addition to several other of the sweet, spicy, and woody note terpenes (eugenol, α-thujene, camphene, d-3-carene, caryophyllene, α-copaene, α-humulene, β-linalool, α- and β-pinene, safrole). Also, the solvents toluene and styrene from the packaging material was associated with cinnamon odor, but they should not contribute significantly to the cinnamon odor due to their higher odor thresholds compared to the terpenes [56,57]. The sour and lemon odors are associated with typical citrus and sour note compounds such as the major citral tr and cis isomers, p-cymene and to some extent acetone and acetic acid. Also, off-odor is associated with these substances, which is difficult to explain. However, D-limonene, which has a citrus like character, is associated with cinnamon. But it must be emphasized, that the odor character of the terpene may vary with concentration, and besides, also masking of some odor compounds may also occur due to interactions and synergistic and antagonistic effects between odor compounds. And their respective odor threshold would need to be taken into consideration to be able to explain the relationship between the sensory odor of the chemical substances. The volatile compounds showed, however, a significant positive correlation (PLSR) with several of the odor attributes: lemon odor ($r = 0.97$, $p < 0.001$, $n = 14$), cinnamon odor ($r = 0.90$, $p < 0.001$, $n = 14$), sulfur odor ($r = 0.86$, $p < 0.001$, $n = 10$), off ($r = 0.88$, $p < 0.001$, $n = 10$), and burnt ($r = 0.74$. $p < 0.001$, $n = 14$).

4. Conclusions

The chicken fillets packaged with active PLA_b materials were preserved as well as fillets packaged with the APET/PE materials, despite 10-fold higher oxygen transmission rate of the PLA_b compared to the APET/PE packages. No negative impact on the drip loss and dry matter content was observed from the active PLA_b composites. The antimicrobial effect of the essential oils was pronounced during the first 8 days of storage. Active PLA_b-based materials especially those with citral inhibited the growth of microorganisms compared to the positive control (PC 9), which is related to the sustained release of active components as confirmed by GC/MS. Furthermore, the TVC results from sampling time 14, 20, and 24 day confirmed that PLA_b-based packaging with PC 3, 5, and 8 are similar in maintaining the microbiological quality of the chicken as compared to the positive control (PC 9). However, relatively high level of bacteria (TVC) was measured already after 14 days of storage in all samples. The release of active volatile compounds (terpenes) from the packaging materials has been confirmed by GC/MS. Sulfur compounds derived from spoilage bacteria flora were also detected and it can be indicated that the additives (citral and cinnamon oil) might have partially masked the negative sulfur odor. This might be a problem; however, further studies are needed to confirm that. The volatile compounds were positively correlated and to a great extent could explain the variation in the sensory odor attributes with significant differences, i.e., lemon, cinnamon, sulfur. High intensities of cinnamon and lemon odor were noticed from the chicken samples packed in active PLA_b composites during sensory analysis which might be a problem with respect to broader consumer acceptance. Future studies will focus on encapsulation of active components and their effect on the product quality.

Supplementary Materials: The following are available online at https://www.mdpi.com/article/10.3390/foods10051126/s1, Table S1. General linear model results, Table S2: Average values and p-values for each sensory attribute—appearance and texture; Table S3: Identified main volatile compounds in the empty packaging and their relative concentrations (area of the chromatographic peak) at the starting, 8th and 14th day of storage. Results are averaged from three parallel samples, Table S4: Detected major volatile compounds in the packaging with chicken samples and their relative concentrations (area of the chromatographic peak) at the 8th and 14th day of storage. Results are averaged from three parallel samples; Figure S1: Dry matter of chicken samples as a function of storage time for different packaging combinations, Figure S2: Bi-plot over samples and attributes—all attributes, O = Odour, T = Texture.

Author Contributions: Conceptualization: J.S. (Jawad Sarfraz) and M.K.P.; Data curation: J.S. (Jawad Sarfraz), J.N., J.S. (Josefine Skaret) and M.K.P.; Formal analysis: J.S. (Jawad Sarfraz), A.Å.H., J.-E.H., T.-A.L., J.N., J.S. (Josefine Skaret), T.P.H. and M.K.P.; Funding acquisition: M.K.P.; Methodology: M.K.P.; Writing—original draft: J.S. (Jawad Sarfraz); Writing—review & editing: J.S. (Jawad Sarfraz), A.Å.H., J.-E.H., J.N., J.S. (Josefine Skaret), T.P.H. and M.K.P. All authors have read and agreed to the published version of the manuscript.

Funding: This research was funded by Foundation for Research Levy on Agricultural Product (Oslo, Norway) through strategic programs FoodMicroPack and FutureFoodControl (grant number 262306, 314743). The APC was funded by Nofima As.

Acknowledgments: The authors would like to thank Tove Maugesten, Anette Wold Åsli, Magnhild Seim Grøvlen and Elin Merete Wetterhus for their skillful technical assistance.

Conflicts of Interest: The authors declare no conflict of interest.

References

1. European Commission. Commission Regulation (EC) No. 450/2009 of 29 May 2009 on active and intelligent materials and articles intended to come into contact with food. *Off. J. Eur. Union.* **2009**, *135*, 3–11.
2. Alparslan, Y.; Baygar, T. Effect of chitosan film coating combined with orange peel essential oil on the shelf life of deepwater pink shrimp. *Food Bioprocess Technol.* **2017**, *10*, 842–853. [CrossRef]
3. Ahmed, I.; Lin, H.; Zou, L.; Brody, A.L.; Li, Z.; Qazi, I.M.; Pavase, T.R.; Lv, L. A comprehensive review on the application of active packaging technologies to muscle foods. *Food Control* **2017**, *82*, 163–178. [CrossRef]
4. Yildirim, S.; Röcker, B.; Pettersen, M.K.; Nilsen-Nygaard, J.; Ayhan, Z.; Rutkaite, R.; Radusin, T.; Suminska, P.; Marcos, B.; Coma, V. Active packaging applications for food. *Compr. Rev. Food Sci. Food Saf.* **2018**, *17*, 165–199. [CrossRef] [PubMed]
5. Mith, H.; Dure, R.; Delcenserie, V.; Zhiri, A.; Daube, G.; Clinquart, A. Antimicrobial activities of commercial essential oils and their components against food-borne pathogens and food spoilage bacteria. *Food Sci. Nutr.* **2014**, *2*, 403–416. [CrossRef]
6. Burt, S. Essential oils: Their antibacterial properties and potential applications in foods—A review. *Int. J. Food Microbiol.* **2004**, *94*, 223–253. [CrossRef]
7. Nazzaro, F.; Fratianni, F.; De Martino, L.; Coppola, R.; De Feo, V. Effect of essential oils on pathogenic bacteria. *Pharmaceuticals* **2013**, *6*, 1451–1474. [CrossRef] [PubMed]
8. Lopez-Romero, J.C.; González-Ríos, H.; Borges, A.; Simões, M. Antibacterial effects and mode of action of selected essential oils components against Escherichia coli and Staphylococcus aureus. *Evid.-Based Complement. Altern. Med.* **2015**, *2015*, 795435. [CrossRef]
9. Bashir, A.; Jabeen, S.; Gull, N.; Islam, A.; Sultan, M.; Ghaffar, A.; Khan, S.M.; Iqbal, S.S.; Jamil, T. Co-concentration effect of silane with natural extract on biodegradable polymeric films for food packaging. *Int. J. Biol. Macromol.* **2018**, *106*, 351–359. [CrossRef]
10. Zhu, J.-Y.; Tang, C.-H.; Yin, S.-W.; Yang, X.-Q. Development and characterization of novel antimicrobial bilayer films based on Polylactic acid (PLA)/Pickering emulsions. *Carbohydr. Polym.* **2018**, *181*, 727–735. [CrossRef]
11. Benbettaïeb, N.; Tanner, C.; Cayot, P.; Karbowiak, T.; Debeaufort, F. Impact of functional properties and release kinetics on antioxidant activity of biopolymer active films and coatings. *Food Chem.* **2018**, *242*, 369–377. [CrossRef]
12. Alarcón-Moyano, J.K.; Bustos, R.O.; Herrera, M.L.; Matiacevich, S.B. Alginate edible films containing microencapsulated lemongrass oil or citral: Effect of encapsulating agent and storage time on physical and antimicrobial properties. *J. Food Sci. Technol.* **2017**, *54*, 2878–2889. [CrossRef]
13. Castro Mayorga, J.L.; Fabra Rovira, M.J.; Cabedo Mas, L.; Sánchez Moragas, G.; Lagaron Cabello, J.M. Antimicrobial nanocomposites and electrospun coatings based on poly(3-hydroxybutyrate-co-3-hydroxyvalerate) and copper oxide nanoparticles for active packaging and coating applications. *J. Appl. Polym. Sci.* **2018**, *135*, 45673. [CrossRef]
14. Higueras, L.; López-Carballo, G.; Hernández-Muñoz, P.; Gavara, R.; Rollini, M. Development of a novel antimicrobial film based on chitosan with LAE (ethyl-Nα-dodecanoyl-L-arginate) and its application to fresh chicken. *Int. J. Food Microbiol.* **2013**, *165*, 339–345. [CrossRef]
15. Wen, P.; Zhu, D.-H.; Feng, K.; Liu, F.-J.; Lou, W.-Y.; Li, N.; Zong, M.-H.; Wu, H. Fabrication of electrospun polylactic acid nanofilm incorporating cinnamon essential oil/β-cyclodextrin inclusion complex for antimicrobial packaging. *Food Chem.* **2016**, *196*, 996–1004. [CrossRef]
16. Rezaei, F.; Shahbazi, Y. Shelf-life extension and quality attributes of sauced silver carp fillet: A comparison among direct addition, edible coating and biodegradable film. *LWT* **2018**, *87*, 122–133. [CrossRef]
17. Sangeetha, V.; Deka, H.; Varghese, T.; Nayak, S. State of the art and future prospectives of poly (lactic acid) based blends and composites. *Polym. Compos.* **2018**, *39*, 81–101. [CrossRef]
18. Radusin, T.; Torres-Giner, S.; Stupar, A.; Ristic, I.; Miletic, A.; Novakovic, A.; Lagaron, J.M. Preparation, characterization and antimicrobial properties of electrospun polylactide films containing *Allium ursinum* L. extract. *Food Packag. Shelf Life* **2019**, *21*, 100357. [CrossRef]
19. Tas, B.A.; Sehit, E.; Tas, C.E.; Unal, S.; Cebeci, F.C.; Menceloglu, Y.Z.; Unal, H. Carvacrol loaded halloysite coatings for antimicrobial food packaging applications. *Food Packag. Shelf Life* **2019**, *20*, 100300.
20. Takma, D.K.; Korel, F. Active packaging films as a carrier of black cumin essential oil: Development and effect on quality and shelf-life of chicken breast meat. *Food Packag. Shelf Life* **2019**, *19*, 210–217. [CrossRef]
21. Nilsen-Nygaard, J.; Fernández, E.N.; Radusin, T.; Rotabakk, B.T.; Sarfraz, J.; Sharmin, N.; Sivertsvik, M.; Sone, I.; Pettersen, M.K. Current status of biobased and biodegradable food packaging materials: Impact on food quality and effect of innovative processing technologies. *Compr. Rev. Food Sci. Food Saf.* **2021**, *20*, 1333–1380. [CrossRef]
22. Ma, Y.; Li, L.; Wang, Y. Development of PLA-PHB-based biodegradable active packaging and its application to salmon. *Packag. Technol. Sci.* **2018**, *31*, 739–746. [CrossRef]
23. Larsen, H.; Kohler, A.; Magnus, E.M. Ambient oxygen ingress rate method—an alternative method to Ox-Tran for measuring oxygen transmission rate of whole packages. *Packag. Technol. Sci. Int. J.* **2000**, *13*, 233–241. [CrossRef]
24. Larsen, H.; Liland, K.H. Determination of O_2 and CO_2 transmission rate of whole packages and single perforations in micro-perforated packages for fruit and vegetables. *J. Food Eng.* **2013**, *119*, 271–276. [CrossRef]
25. Pettersen, M.K.; Grøvlen, M.S.; Evje, N.; Radusin, T. Recyclable mono materials for packaging of fresh chicken fillets: New design for recycling in circular economy. *Packag. Technol. Sci.* **2020**, *33*, 485–498. [CrossRef]

26. Zhu, G.; Feng, N.; Xiao, Z.; Zhou, R.; Niu, Y. Production and pyrolysis characteristics of citral–monochlorotriazinyl-β-cyclodextrin inclusion complex. *J. Therm. Anal. Calorim.* **2015**, *120*, 1811–1817. [CrossRef]
27. Tian, H.; Lu, Z.; Li, D.; Hu, J. Preparation and characterization of citral-loaded solid lipid nanoparticles. *Food Chem.* **2018**, *248*, 78–85. [CrossRef] [PubMed]
28. Silverstein, R.M. *Spectrometric Identification of Organic Compounds*; Silverstein, R.M., Bassler, G.C., Morrill, T.C., Eds.; Wiley & Sons: Hoboken, NJ, USA, 1974.
29. Zeid, A.; Karabagias, I.K.; Nassif, M.; Kontominas, M.G. Preparation and evaluation of antioxidant packaging films made of polylactic acid containing thyme, rosemary, and oregano essential oils. *J. Food Process. Preserv.* **2019**, *43*, 14102. [CrossRef]
30. Hamad, K.; Kaseem, M.; Yang, H.; Deri, F.; Ko, Y. Properties and medical applications of polylactic acid: A review. *Express Polym. Lett.* **2015**, *9*, 435–455. [CrossRef]
31. Panseri, S.; Martino, P.; Cagnardi, P.; Celano, G.; Tedesco, D.; Castrica, M.; Balzaretti, C.; Chiesa, L. Feasibility of biodegradable based packaging used for red meat storage during shelf-life: A pilot study. *Food Chem.* **2018**, *249*, 22–29. [CrossRef]
32. Pettersen, M.; Bardet, S.; Nilsen, J.; Fredriksen, S. Evaluation and suitability of biomaterials for modified atmosphere packaging of fresh salmon fillets. *Packag. Technol. Sci.* **2011**, *24*, 237–248. [CrossRef]
33. Jiménez, S.; Salsi, M.; Tiburzi, M.; Rafaghelli, R.; Tessi, M.; Coutaz, V. Spoilage microflora in fresh chicken breast stored at 4 C: Influence of packaging methods. *J. Appl. Microbiol.* **1997**, *83*, 613–618. [CrossRef]
34. Pettersen, M.K.; Hansen, A.Å.; Mielnik, M. Effect of different packaging methods on quality and shelf life of fresh reindeer meat. *Packag. Technol. Sci.* **2014**, *27*, 987–997. [CrossRef]
35. Van Moeseke, W.; De Smet, S. Effect of time of deboning and sample size on drip loss of pork. *Meat Sci.* **1999**, *52*, 151–156. [CrossRef]
36. Jakobsen, M.; Bertelsen, G. The use of CO_2 in packaging of fresh red meats and its effect on chemical quality changes in the meat: A review. *J. Muscle Foods* **2002**, *13*, 143–168. [CrossRef]
37. Patsias, A.; Badeka, A.; Savvaidis, I.; Kontominas, M. Combined effect of freeze chilling and MAP on quality parameters of raw chicken fillets. *Food Microbiol.* **2008**, *25*, 575–581. [CrossRef]
38. Holck, A.L.; Pettersen, M.K.; Moen, M.H.; Sørheim, O. Prolonged shelf life and reduced drip loss of chicken filets by the use of carbon dioxide emitters and modified atmosphere packaging. *J. Food Prot.* **2014**, *77*, 1133–1141. [CrossRef] [PubMed]
39. Rotabakk, B.T.; Birkeland, S.; Jeksrud, W.K.; Sivertsvik, M. Effect of modified atmosphere packaging and soluble gas stabilization on the shelf life of skinless chicken breast fillets. *J. Food Sci.* **2006**, *71*, S124–S131. [CrossRef]
40. Chan, S.S.; Roth, B.; Skare, M.; Hernar, M.; Jessen, F.; Løvdal, T.; Jakobsen, A.N.; Lerfall, J. Effect of chilling technologies on water holding properties and other quality parameters throughout the whole value chain: From whole fish to cold-smoked fillets of Atlantic salmon (*Salmo salar*). *Aquaculture* **2020**, *526*, 735381. [CrossRef]
41. Yin, H.; Gilbert, E.R.; Chen, S.; Wang, Y.; Zhang, Z.; Zhao, X.; Zhang, Y.; Zhu, Q. Effect of hybridization on carcass traits and meat quality of erlang mountainous chickens. *Asian-Australas. J. Anim. Sci.* **2013**, *26*, 1504. [CrossRef]
42. Shaarani, S.M.; Nott, K.P.; Hall, L.D. Combination of NMR and MRI quantitation of moisture and structure changes for convection cooking of fresh chicken meat. *Meat Sci.* **2006**, *72*, 398–403. [CrossRef]
43. Borch, E.; Kant-Muermans, M.-L.; Blixt, Y. Bacterial spoilage of meat and cured meat products. *Int. J. Food Microbiol.* **1996**, *33*, 103–120. [CrossRef]
44. Granum, P. Bacillus species. In *The Microbiological Safety and Quality of Food*; Lund, B.M., Baird-Parker, T.C., Gould, G.W., Eds.; Aspen Publishers: New York, NY, USA, 2000; Volume 1, pp. 214–234.
45. García-Soto, B.; Miranda, J.M.; Rodríguez-Bernaldo de Quirós, A.; Sendón, R.; Rodríguez-Martínez, A.V.; Barros-Velázquez, J.; Aubourg, S.P. Effect of biodegradable film (lyophilised alga *Fucus spiralis* and sorbic acid) on quality properties of refrigerated megrim (*Lepidorhombus whiffiagonis*). *Int. J. Food Sci. Technol.* **2015**, *50*, 1891–1900. [CrossRef]
46. Pettersen, M.; Nissen, H.; Eie, T.; Nilsson, A. Effect of packaging materials and storage conditions on bacterial growth, off-odour, pH and colour in chicken breast fillets. *Packag. Technol. Sci. Int. J.* **2004**, *17*, 165–174. [CrossRef]
47. Vihavainen, E.; Lundström, H.-S.; Susiluoto, T.; Koort, J.; Paulin, L.; Auvinen, P.; Björkroth, K.J. Role of broiler carcasses and processing plant air in contamination of modified-atmosphere-packaged broiler products with psychrotrophic lactic acid bacteria. *Appl. Environ. Microbiol.* **2007**, *73*, 1136–1145. [CrossRef]
48. Senanayake, U.M.; Lee, T.H.; Wills, R.B. Volatile constituents of cinnamon (*Cinnamomum zeylanicum*) oils. *J. Agric. Food Chem.* **1978**, *26*, 822–824. [CrossRef]
49. Eilamo, M.; Kinnunen, A.; Latva-Kala, K.; Ahvenainen, R. Effects of packaging and storage conditions on volatile compounds in gas-packed poultry meat. *Food Addit. Contam.* **1998**, *15*, 217–228. [CrossRef]
50. Nychas, G.-J.E.; Skandamis, P.N.; Tassou, C.C.; Koutsoumanis, K.P. Meat spoilage during distribution. *Meat Sci.* **2008**, *78*, 77–89. [CrossRef]
51. Gram, L.; Ravn, L.; Rasch, M.; Bruhn, J.B.; Christensen, A.B.; Givskov, M. Food spoilage—Interactions between food spoilage bacteria. *Int. J. Food Microbiol.* **2002**, *78*, 79–97. [CrossRef]
52. Mikš-Krajnik, M.; Yoon, Y.-J.; Yuk, H.-G. Detection of volatile organic compounds as markers of chicken breast spoilage using HS-SPME-GC/MS-FASST. *Food Sci. Biotechnol.* **2015**, *24*, 361–372. [CrossRef]
53. Lambert, A.D.; Smith, J.P.; Dodds, K.L. Shelf life extension and microbiological safety of fresh meat—A review. *Food Microbiol.* **1991**, *8*, 267–297. [CrossRef]

54. Tománková, J.; Bořilová, J.; Steinhauserová, I.; Gallas, L. Volatile organic compounds as biomarkers of the freshness of poultry meat packaged in a modified atmosphere. *Czech J. Food Sci.* **2012**, *30*, 395–403. [CrossRef]
55. McLeod, A.; Hovde Liland, K.; Haugen, J.E.; Sørheim, O.; Myhrer, K.S.; Holck, A.L. Chicken fillets subjected to UV-C and pulsed UV light: Reduction of pathogenic and spoilage bacteria, and changes in sensory quality. *J. Food Saf.* **2018**, *38*, e12421. [CrossRef] [PubMed]
56. Leonardos, G.; Kendall, D.; Barnard, N. Odor threshold determination of 53 odorant chemicals. *J. Environ. Conserv. Eng.* **1974**, *3*, 579–585. [CrossRef]
57. Cometto-Muñiz, J.E.; Cain, W.S.; Abraham, M.H.; Kumarsingh, R. Sensory Properties of Selected Terpenes: Thresholds for Odor, Nasal Pungency, Nasal Localization, and Eye Irritationa. *Ann. N. Y. Acad. Sci.* **1998**, *855*, 648–651. [CrossRef]

Article

Efficacy of Biopolymer/Starch Based Antimicrobial Packaging for Chicken Breast Fillets

Noor L. Yusof [1], Noor-Azira Abdul Mutalib [1], U. K. Nazatul [1], A. H. Nadrah [1], Nurain Aziman [2], Hassan Fouad [3], Mohammad Jawaid [4,*], Asgar Ali [5], Lau Kia Kian [4] and Mohini Sain [6]

1 Faculty of Food Science and Technology, Universiti Putra Malaysia, Serdang 43400, Selangor, Malaysia; noorliyana@upm.edu.my (N.L.Y.); n_azira@upm.edu.my (N.-A.A.M.); nazatulumira.karim@gmail.com (U.K.N.); nadrahabdhalid@gmail.com (A.H.N.)
2 Alliance of Research & Innovation for Food (ARIF), Faculty of Applied Sciences, Universiti Teknologi MARA, Cawangan Negeri Sembilan, Kampus Kuala Pilah, Kuala Pilah 72000, Negeri Sembilan, Malaysia; ainaziman@uitm.edu.my
3 Applied Medical Science Department, Community College, King Saud University, P.O. Box 10219, Riyadh 11433, Saudi Arabia; menhfef@ksu.edu.sa
4 Laboratory of Biocomposite Technology, Institute of Tropical Forestry and Forest Products (INTROP), Universiti Putra Malaysia, Serdang 43400, Selangor, Malaysia; laukiakian@gmail.com
5 Centre of Excellence for Postharvest Biotechnology, School of Biosciences, University of Notthingham Malaysia, Semenyih 43500, Selangor, Malaysia; asgar.ali@nottingham.edu.my
6 Centre for Biocomposites and Biomaterials Processing, University of Toronto, 33 Willcocks Street, Toronto, ON M5S3B3, Canada; m.sain@utoronto.ca
* Correspondence: jawaid@upm.edu.my

Citation: Yusof, N.L.; Mutalib, N.-A.A.; Nazatul, U.K.; Nadrah, A.H.; Aziman, N.; Fouad, H.; Jawaid, M.; Ali, A.; Kian, L.K.; Sain, M. Efficacy of Biopolymer/Starch Based Antimicrobial Packaging for Chicken Breast Fillets. *Foods* **2021**, *10*, 2379. https://doi.org/10.3390/foods10102379

Academic Editors: Valeria Rizzo and Muratore Giuseppe

Received: 31 July 2021
Accepted: 29 August 2021
Published: 8 October 2021

Publisher's Note: MDPI stays neutral with regard to jurisdictional claims in published maps and institutional affiliations.

Abstract: Food contamination leading to the spoilage and growth of undesirable bacteria, which can occur at any stage along the food chain, is a significant problem in the food industry. In the present work, biopolymer polybutylene succinate (PBS) and polybutylene succinate/tapioca starch (PBS/TPS) films incorporating Biomaster-silver (BM) and SANAFOR® (SAN) were prepared and tested as food packaging to improve the lifespan of fresh chicken breast fillets when kept in a chiller for seven days. The incorporation of BM and SAN into both films demonstrated antimicrobial activity and could prolong the storability of chicken breast fillets until day 7. However, PBS + SAN 2%, PBS/TPS + SAN 1%, and PBS/TPS + SAN 2% films showed the lowest microbial log growth. In quality assessment, incorporation of BM and SAN into both film types enhanced the quality of the chicken breast fillets. However, PBS + SAN 1% film showed the most notable enhancement of chicken breast fillet quality, as it minimized color variation, slowed pH increment, decreased weight loss, and decelerated the hardening process of the chicken breast fillets. Therefore, we suggest that the PBS + SAN and PBS/TPS + SAN films produced in this work have potential use as antimicrobial packaging in the future.

Keywords: biomaster-silver; SANAFOR®; tapioca starch; polybutylene succinate; antimicrobial; food packaging

1. Introduction

Over the years, the food industry has faced huge food waste, especially with perishable foods. The whole process of food contamination leads to food spoilage, consequently leading to limited shelf life and low product quality. This can be seen through the changes in texture, color, and nutritive value, as well as microbial growth [1]. The storability of chicken fillets is often short term owing to their susceptibility to microbial spoilage [2]. High protein and moisture contents in the meat serve as a competent substrate for microbial growth. With the presence of oxygen, the lipid content of the chicken fillet enhances the oxidation reaction, further deteriorating the quality of the meat [3]. Thus, it depicts a higher risk for human consumption and economic loss for producers [4].

To prolong the lifespan of chicken-based foodstuffs, manufacturers have opted to wrap and package the fresh product using different types of packaging, mainly biopolymer. Biopolymer is used as an alternative to plastic packaging in producing environmentally friendly packaging. Besides reducing the accumulation of plastic waste, the use of biopolymers can also lessen the utilization of industrial agro- and biomass waste-derived plastics [5]. Polybutylene succinate (PBS) is a polymer with biocompatible characteristics that can be degraded by bacteria and fungi in landfills or the sea [6]. The mechanical and physical properties of PBS are comparable to polypropylene (PP) and polyethylene (PE), which are widely applied by the food packaging industry [7]. PBS also has the potential to replace polyolefin in the future [8]. Starch has also been considered as a polymer with biodegradable criteria besides its abundance and accessibility at a low cost. These advantages have given it great potential for packaging purposes in various fields. One of the most analyzed starch-based polymers is thermoplastic starch, which is widely applied in food packaging owing to its simple processing by common equipment in the plastic manufacturing industries, such as casting, extrusion, blow film extrusion, injection molding, and compression molding [9]. Moreover, pre-plasticized starch could promote larger surface reaction and improve homogenization during compounding thermoplastic starch/PBS thin films [10]. The addition of maleated PBS as compatibilizer to thermoplastic starch/PBS blends could provide the combined properties of great water resistance, enhanced strength, and good biodegradability, and is anticipated to serve as promising packing material [11]. However, the application of PBS is restricted due to several limitations, such as microbiological corrosion and low mechanical strength [12]. Besides that, some of the disadvantages that limit the wide use of thermoplastic starch in packaging are attributed to its self-deteriorating behavior over time as well as its high sensitivity to moisture, resulting in starch retrogradation [8]. Hence, a lot of research on different antimicrobial additives and natural fillers has been done to widen the functionality of starch-based materials [5] and PBS films.

Active packaging is a cutting-edge idea in which the package, the product, and the environment work together to extend the shelf life of foodstuffs while also improving sensory and safety characteristics, thereby retaining product quality. Antimicrobial food packaging is a type of active packaging that incorporates antimicrobial substances in food packaging to reduce pathogenic microorganisms and eliminate unpleasant changes in food quality, which in turn enhances the product's shelf life [1,13]. Several studies on the effectiveness of antimicrobial packaging on food products have been done by previous researchers. Zhao et al. [14] produced soy protein isolation/antimicrobial silver nanoparticle films showing antibacterial activity that was efficient against Gram-positive and Gram-negative bacteria. After 28 days in a 4–10 °C temperature range, the population of Listeria monocytogenes on turkey frankfurters coated with soy protein film containing antibacterial nisin blended with either green tea extract (1%) or grape seed extract (1%) had fallen by more than 2 log cycles [15]. The bacterial flora in milk and *Micrococcus luteus* ATCC 10,240 were both effectively inhibited by the nisin-coated films [16]. Incorporation of grape seed extract also largely improved the puncture, thickness, and tensile strengths of a soy protein isolate/nisin/ethylenediaminetetraacetic acid film, and it also demonstrated the greatest inhibitory activity against *Escherichia coli* O157:H7, *Listeria monocytogenes*, and *Salmonella typhimurium* populations [17]. Another reported study fabricated ethylene–vinyl alcohol copolymer novel films mixed with cocoa extract in 10%, 15%, and 20% amounts, which presented great bactericidal activity against *Salmonella enterica*, *Escherichia coli*, *Listeria monocytogenes*, and *Staphylococcus aureus* [18]. In addition, some studies reported that the constituents of plant-derived essential oils such as lemongrass oil, oregano oil, cinnamon oil, carvacrol, cinnamaldehyde, and citral could be utilized to prepare edible-based antimicrobial films for packaging food products [19].

Recently, the development of novel antimicrobial agents against bacteria through advanced technology has been regarded as a prioritized area in the field of biomedical research. Biomaster silver (BM) is a silver-based antimicrobial material on the market. BM is based on silver ion technology, and Llorens et al. [13] mentioned that silver ions are known to possess the highest antimicrobial capacity of the metallic cations and is able to inhibit a wide range of microbial organisms. It possesses low vitality and long-term biocide properties, yet a low level of toxicity to eukaryotic cells. The interaction of silver ions with the ribosomes would inhibit the enzyme expression and interfere with the permeability of the membrane. It interacts with nucleic acids and cytoplasmatic components, and alters the enzyme mechanism behavior after binding with protein thiol groups [20]. According to Xu et al. [21], previous studies have shown that an increase in reactive oxygen species (ROS) was induced by the presence of metal nanoparticles, posing toxic effects correlated with oxidative stress. The inhibition of respiratory enzymes was caused by silver ions that interacted with the thiol group of the enzymes, thus producing ROS. It is considered one of the critical factors that can affect oxidative stress, as ROS is able to react directly with DNA and proteins or with lipids in order to induce the production of malondialdehyde, which can react with DNA, proteins, and other lipids [22]. A higher temperature has also been proven to induce higher antibacterial activity in silver nanoparticles through the increase of the ROS formation level [21]. Apart from that, SANAFOR® (SAN), a commercialized antimicrobial agent, controls microbial growth on plastic surfaces, protecting against bacteria, fungi, and algae that may degrade the food products. Although the active substances used in the antimicrobial agent were not mentioned, it was said that SAN could provide good thermal stability with low water solubility that may limit the migration from plastic.

The goal of this work is to address the various modifications and preparation methodologies of preparing BM and SAN as promising materials that can be applies as coating or packaging material in food, providing new insights into the numerous applications to maintain the quality and safety of the food products contained in the packaging for a specific time period. Although BM and SAN are not newly developed antimicrobial agents, the incorporation of both compounds has not been fully explored when it comes to its application in food packaging, especially for a perishable commodity like poultry. There are no available studies focusing on incorporating polybutylene succinate (PBS) and polybutylene succinate/tapioca starch (PBS/TPS) with BM and SAN per se, and these materials exhibited strong antimicrobial activity against microorganisms, especially in poultry products. Hence, the idea of incorporating BM and SAN into PBS/TPS film could open up new possibilities of developing a sustainable packaging film for preserving a few other perishable foods. The novelty of this study is that the modification of BM and SAN with different contents was done to tailor the optimum concentration that can be useful in food packaging, especially to ensure the microbiological safety and maintain the physicochemical quality of chicken breast fillets during storage, as well as its environmentally friendly materials. Antimicrobial packaging, on the other hand, is still a difficult technology to master, with only a few items commercialized on the market. Furthermore, based on our knowledge, no study has been conducted regarding the incorporation of BM and SAN into PBS and PBS/TPS films. Therefore, the present work aims to investigate the effects of PBS and PBS/TPS films incorporated with BM and SAN on the shelf life of fresh chicken breast fillets. Therefore, the novelty of this study is the generation of antimicrobial packaging films from PBS and PBS/TPS materials for food preservation purposes.

2. Materials and Methods

2.1. Materials

PBS pellets (BioPBS™ FZ71PB) (density: 1.26 g/cm^3) were purchased from PTT Public Company Limited in Bangkok, Thailand. The powder form of tapioca starch (TPS) was bought from PT Starch Solution in Karawang, Indonesia. Meanwhile, Biomaster-silver (BM) and SANAFOR® (SAN) were purchased from Malaysian companies. Other materials like agar plate and peptone water were procured from Oxoid, Malaysia. Fresh chicken breast fillets were obtained from a supply chain company, Segi Fresh, Balakong, Selangor.

2.2. Fabrication of PBS and PBS/TPS Films

In this study, PBS and PBS/TPS films were prepared through the melting-blow method since this technique can facilitate the dispersion of different added components with the PBS matrix. Two distinct antimicrobial substances, namely, BM and SAN, were incorporated into two different films, (i) PBS and (ii) PBS, with the addition of TPS. The antimicrobial substances were incorporated at different concentrations. Two untreated films, PBS film and PBS film, were prepared with TPS (PBS/TPS) added. A total of 10 films were provided, as shown in Table 1. The films were manually laminated onto a kraft paper tray (11.5 × 9 × 5 cm) aseptically. As for the control, a polypropylene (PP) microwavable container was used.

Table 1. A total of 10 formulations of PBS films.

No.	PBS Film Code	Antimicrobial Agent	g/g (%) Amount of Antimicrobial Agent
1	PBS (untreated)	N/A	N/A
2	PBS + BM 1.5%	BM	1.5
3	PBS + BM 3%	BM	3
4	PBS + SAN 1%	SAN	1
5	PBS + SAN 2%	SAN	2
4	PBS/TPS (untreated)	N/A	N/A
5	PBS/TPS + BM 1.5%	BM	1.5
6	PBS/TPS + BM 3%	BM	3
7	PBS/TPS + SAN 1%	SAN	1
8	PBS/TPS + SAN 2%	SAN	2

Note: N/A: not available.

2.3. Sampling and Storage

The chicken breast fillet samples, weighing approximately 30 ± 3 g, were placed on the laminated kraft paper tray. The trays were covered with the same lamination film, then kept for 7 days at 4 °C. The procedure was carried out in triplicate for each sample. Microbiological analysis and quality assessment were conducted on days 0, 1, 3, 5, and 7 during storage. The analyses on the initial day (day 0) were carried out upon freshly bought chicken fillets with no treatment applied.

2.4. Microbiological Analysis

Microbiological analysis was determined by total plate count (TPC) according to CLSI (2010). The chicken breast fillets were cut into smaller pieces with a weight of around 25 g. Each sample was then moved to stomacher bags that were filled with 225 mL of 0.1% sterilized peptone solution, followed by 60 s homogenization within the stomacher (Tekma Lab Blender 80, Seward Medical, UK). Suitable serial dilution was then applied to the peptone water (0.1%) contained in each sample. Afterwards, about 0.1 mL of the diluted homogenates were evenly distributed on Plate Count Agar (Oxoid, UK) using an L-shaped rod. Then, the inoculated plates underwent incubation for 18–24 h at 37 °C. The TPC was counted for each sample as log CFU/g, representing the logarithms of colony-forming units per gram [4].

2.5. Quality Assessment

2.5.1. pH Analysis

The pH values of the chicken samples were examined following the AOAC method. Approximately 10 g sample were mixed with 100 mL distilled water and then the filtrate was collected for pH measurement using a pH meter (Mettler-Toledo International Inc., Columbus, OH, USA).

2.5.2. Weight Loss

The changes in the weight of the chicken breast fillet samples were recorded on days 1, 3, 5, and 7 upon storage. The weight loss was presented as lost percentage in respect to the initial weight of the samples [4].

2.5.3. Texture Analysis

The texture of the chicken breast fillet samples was analyzed with a texture analyzer (TA.HDplusC, Stable Micro Systems Ltd., Godalming, UK), and the texture was described in terms of hardness. Three different positions were located for each sample with perpendicular measurements to the chicken breast fillet surface, whereas the mean values were analyzed in triplicate measurement from the fillet samples to obtain an average value of the hardness [23].

2.5.4. Color Analysis

The measurement of color for the chicken breast fillets was performed with a chroma meter (CR-410, Konica Minolta, Japan). Values of a* (redness), b* (yellowness), and L* (lightness) were used to characterize color. Three different positions were located for each sample with perpendicular measurements to the chicken breast fillet surface [24]. The mean values (a*, b*, and L*) were analyzed in triplicate measurement from the fillet samples to obtain an average value of the hardness [25].

2.5.5. Overall Visual Quality

The overall observation of the chicken breast fillet samples throughout the storage was conducted with a camera (Apple iPhone 6 Plus, Apple Inc., Cupertino, CA, USA).

2.6. Statistical Analysis

Triplicate work was conducted for all analyses in this study, and the collected data were analyzed by one-way ANOVA to obtain the values of mean ± standard deviation. Meanwhile, the statistical differences were regarded as significant as if $p < 0.05$ (Duncan's multiple range test). All statistical evaluations for the results were performed using Minitab 18.0 Statistical software.

3. Results

3.1. Total Plate Count (TPC)

The TPC of the chicken breast fillet samples stored in the PP container and PBS film laminated tray during storage at 4 °C for seven days is shown in Figure 1. Meanwhile, the TPC of the chicken breast fillet samples stored in the PP container and PBS/TPS film laminated tray during storage at 4 °C for seven days is shown in Figure 2.

The initial TPC of the chicken samples without any treatment (day 0) was 4.17 log CFU/g. Based on Figure 1, the log CFU/g value increased in all treatments during storage at 4 °C up to seven days, with values between 4.17 and 7.41 log CFU/g. The limit of acceptability for meat product is 7.00 log CFU/g [26]. Among all samples, the untreated PBS and PBS + BM 1.5% exceeded the acceptability limit on the third day of storage. However, the PP container (control) surpassed the acceptability limit on the fifth day. The untreated PBS and PBS + BM 1.5% films showed higher TPC than the control throughout the seven days, as PBS showed low gas barrier properties [27] that may induce microbial growth. The incorporation of 3% BM and 1% SAN into the PBS improved the film antimicrobial activity, and the lifespan of chicken samples stored in PBS + BM 3% and PBS + SAN 1% films could be extended up to seven days. Meanwhile, the chicken sample stored in the PBS + SAN 2% film showed the significantly lowest ($p < 0.05$) TPC and was still acceptable on day 7, indicating that 2% SAN exhibits the highest antimicrobial properties compared to the others.

As illustrated in Figure 2, the log CFU/g value increased in all treatments during storage at 4 °C up to seven days with values between 4.17 and7.25 log CFU/g. The addition of TPS on both untreated and treated films resulted in lower log values than the control from day 3 storage onwards. This further indicates that the addition of TPS to PBS made it partially compatible polymer blends [11]. Among all the samples, the PP container (control) exceeded the acceptability limit on the fifth day of storage; however, other treatments for chicken samples were still acceptable up to seven days. Among the treated films, the incorporation of 1% and 2% SAN into PBS/TPS enhanced the antimicrobial activity by exhibiting the significantly lowest ($p < 0.05$) log value compared to the others up to seven days.

Appendini and Hotchkiss [28] stated that the introduction of antimicrobials into the packaging is to avoid surface growth on foods, such as the spoilage that may occur primarily on the surface of intact meat. The antimicrobials were gradually released from the packaging film, which may be better compared to the dipping and spraying technique. The techniques of directly adding preservatives may cause rapid diffusion of the antibacterial agent into food and denaturize its active sites, which ultimately lowers the reactivity with bacteria. However, antimicrobial-incorporated packaging allows antimicrobial agents to migrate slowly and continuously from the container to the food surface, which helps enhance the high concentration of antimicrobial agents over a longer period [2]. They also discovered that as the storage temperature rises, the movement of active chemicals from film to food accelerates. We did not observe this effect in the present study since our chicken samples were stored at 4 °C, and the interactions between coating materials, target bacteria, antimicrobial agents, and the food matrix themselves could differ, therefore influencing active compound release rates.

As reported by Warsiki and Bawardi [29], antibacterial packaging materials could be made from tapioca starch films containing antimicrobial ZnO nanoparticles. After integration with 2% ZnO, the film showed an inhibition index of around 7.67 mm against E. coli. PBS with 10 wt.% thymol was also effective for inhibiting E. coli growth [12]. Cardoso et al. [30] tested the antimicrobial efficiency of oregano essential oil (OEO)-filled poly (butylene adipate co-terephthalate) active films for fish fillet preservation at 7 °C. All formulations except 2.5 g OEO showed improved fillet shelf life reaching up to 10 days. The polylactic acid-based composite films containing 50% cinnamon oil presented antimicrobial behavior against Salmonella Typhimurium and Listeria monocytogenes when inoculated in chicken samples for 16 days of storage in refrigerated conditions [31]. Chitosan films incorporated with 2% ethanolic propolis extract and 1–2% cellulose nanoparticle was also a viable option in slowing microbial development as well as protein and lipid oxidation in minced beef meat [32].

In this study, the antimicrobial agents BM and SAN were effective in PBS and PBS/TPS films. The antimicrobial mechanism of BM silver ion technology is based on the release of silver ions into the moisture layer that naturally exists on a product's surface, then permeates through bacterial cell walls, deactivating essential energy-producing metabolic enzymes and stopping bacteria from growing [33]. However, the mechanism by which SAN inhibits bacteria growth has not been elucidated.

Overall, the log value of the PBS/TPS films with antimicrobial agents was lower than in the PBS films with antimicrobial agents. Furthermore, SAN was shown to have lower microbial growth and was more effective than BM in both PBS and PBS/TPS films. This study concluded that PBS with 2% SAN and PBS/TPS with 1% SAN or 2% SAN were more effective in serving as antimicrobial packaging.

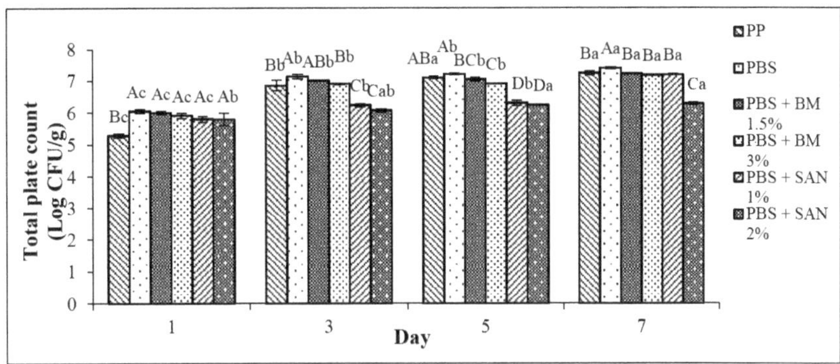

Figure 1. Total plate count (TPC) of chicken breast fillets stored in PP container and PBS film laminated tray during storage at 4 °C for seven days. Note: PP: polypropylene microwavable container (control); PBS: PBS film (untreated); PBS + BM 1.5%: PBS film + 1.5% Biomaster silver; PBS + BM 3%: PBS film + 3% Biomaster silver; PBS + SAN 1%: PBS film + 1% SANAFOR; PBS + SAN 2%: PBS film + 2% SANAFOR. Error bars indicate standard deviation (n = 3). The different A–D capital letters are significantly different ($p < 0.05$) among treatment for each day. The different a–c lowercase letters are significantly different ($p < 0.05$) among storage day for each sample.

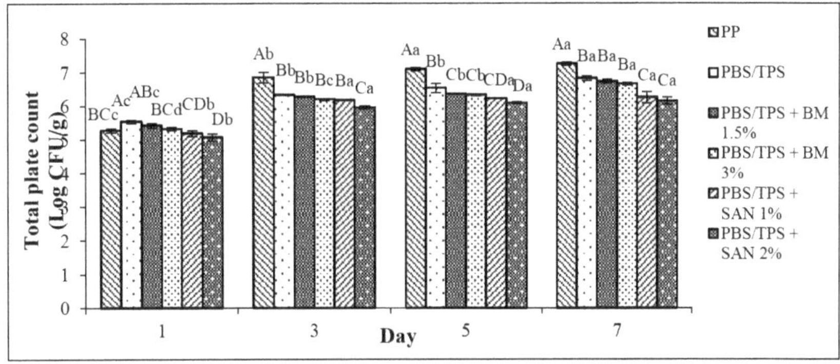

Figure 2. Total plate count (TPC) of chicken breast fillets stored in PP container and PBS/TPS film laminated tray during storage at 4 °C for seven days. Note: PP: polypropylene microwavable container (control); PBS/TPS: PBS/TPS film (untreated); PBS/TPS + BM 1.5%: PBS/TPS film + 1.5% Biomaster silver; PBS/TPS + BM 3%: PBS/TPS film + 3% Biomaster silver; PBS/TPS + SAN 1%: PBS/TPS film + 1% SANAFOR; PBS/TPS + SAN 2%: PBS/TPS film + 2% SANAFOR. Error bars indicate standard deviation (n = 3). The different A–D capital letters are significantly different ($p < 0.05$) among treatment for each day. The different a–d lowercase letters are significantly different ($p < 0.05$) among storage day for each sample.

3.2. pH Value

The pH changes in value for the chicken samples stored in PBS and PBS/TPS film laminated tray during storage at 4 °C for seven days are shown in Figures 3 and 4, respectively. The pH value shown by the chicken samples initially in this study without any treatment (day 0) was 6.15 ± 0.06.

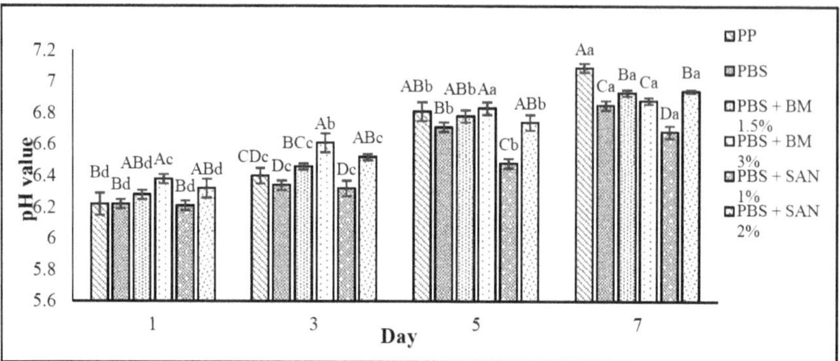

Figure 3. The pH values of chicken breast fillets stored in PP container and PBS film laminated tray during storage at 4 °C for seven days. Note: PP: polypropylene microwavable container (control); PBS: PBS film (untreated); PBS + BM 1.5%: PBS film + 1.5% Biomaster silver; PBS + BM 3%: PBS film + 3% Biomaster silver; PBS + SAN 1%: PBS film + 1% SANAFOR; PBS + SAN 2%: PBS film + 2% SANAFOR. Error bars indicate standard deviation (n = 3). The different A–F capital letters are significantly different ($p < 0.05$) among treatment for each day. The different a–d lowercase letters are significantly different ($p < 0.05$) among storage day for each sample.

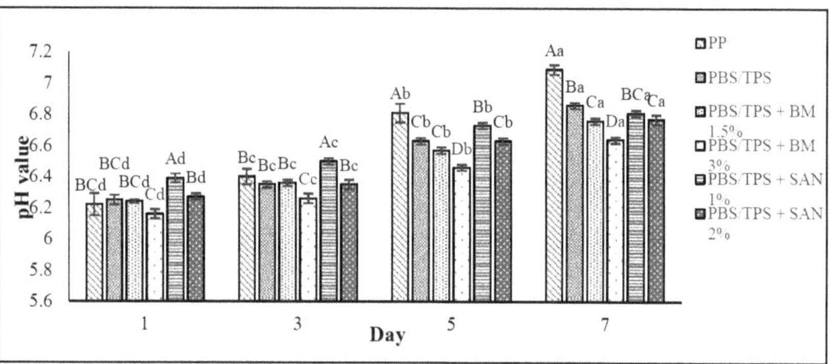

Figure 4. The pH values of chicken breast fillets stored in PP container and PBS/TPS film laminated tray during storage at 4 °C for seven days. Note: PP: polypropylene microwavable container (control); PBS/TPS: PBS/TPS film (untreated); PBS/TPS + BM 1.5%: PBS/TPS film + 1.5% Biomaster silver; PBS/TPS + BM 3%: PBS/TPS film + 3% Biomaster silver; PBS/TPS + SAN 1%: PBS/TPS film + 1% SANAFOR; PBS/TPS + SAN 2%: PBS/TPS film + 2% SANAFOR. Error bars indicate standard deviation (n = 3). The different A–D capital letters are significantly different ($p < 0.05$) among treatment for each day. The different a–d lowercase letters are significantly different ($p < 0.05$) among storage day for each sample.

Our findings show that as the storage period increased, the pH value of the chicken samples increased. However, the pH values of the chicken samples in PBS and PBS/TPS films incorporating BM and SAN were slightly increased compared to the control sample. The chicken samples stored in PP container (control) showed the highest pH value (7.09 ± 0.03) at the end of storage, and a pH value above 7 is considered to indicate negative sensory attributes [34].

Similar trends of pH increment throughout the storage period of chicken and beef were also observed in recent studies by Rashidaie et al. [35] and Katiyo et al. [36]. The accumulated alkali nitrogenous components such as amines and ammonia produced by either microbial or endogenous enzymes could cause an increased pH value, which subsequently leads to the meat color darkening [36,37]. Therefore, the low pH increment of chicken samples stored in the treated films in this study might be contributed to the low color variation of chicken samples, as discussed earlier in this study. The slow pH increment in chicken samples stored in treated films might be due to the ability of antimicrobial agents BM and SAN to prohibit enzyme activity [37] and low oxidation rates occurred during storage [34]. When compared to unpackaged poultry, samples packed with chitosan/montmorillonite bionanocomposites incorporating ginger essential oil showed a 1.2 to 2.6 log CFU/g reduction in microbial count and were able to maintain their color and pH values [38]. Therefore, the pH values of chicken samples were correlated with the total plate count and color. Overall, the lowest pH values were obtained for the chicken breast fillets samples stored in PBS + SAN 1% and PBS/TPS + BM 3% films.

3.3. Weight Loss

The weight loss of the chicken breast fillet samples stored in the PBS and PBS/TPS film laminated tray during storage at 4 °C for seven days is shown in Figures 5 and 6, respectively. Based on both figures, the weight loss increased ($p < 0.05$) throughout seven days of the storage period for all samples, with values between 3.18 and 6.83% (Figure 5) and 2.83 and 6.83% (Figure 6).

The weight loss of the chicken samples in PBS and PBS/TPS film incorporating BM and SAN exhibited a slight increase throughout the period of storage compared to the control sample. The results show that chicken samples stored in PBS + SAN 1% and PBS + BM 1.5% film (Figure 5) and PBS/TPS + BM 3% film (Figure 6) recorded the lowest weight loss on day 7 (5.52 ± 0.04%, 5.95 ± 0.02%, and 5.78 ± 0.03%, respectively) compared to the chicken samples stored in the PP container (control) and other films. The increase in weight loss throughout storage was due to the moisture loss from the chicken breast fillets and further resulted in the hardening of the meat texture [34]. As reported by Amjadi et al. [39], the chicken fillets wrapped with ZnO nanoparticles and chitosan nanofiber-filled gelatin-based nanocomposite exhibited less weight loss than the control at the end of the 12th day of storage.

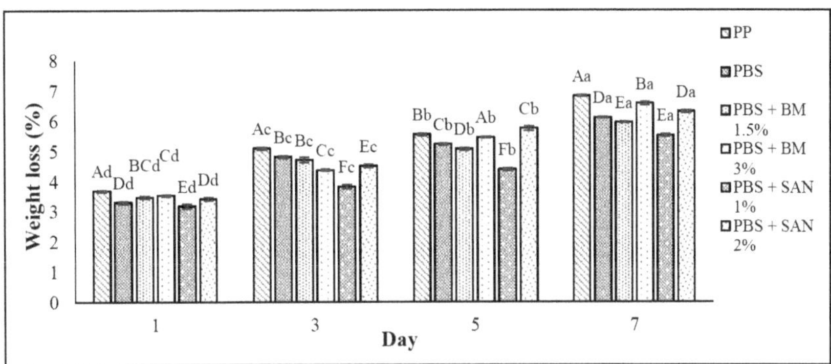

Figure 5. The weight loss values of chicken breast fillets stored in PP container and PBS film laminated tray during storage at 4 °C for seven days. Note: PP: polypropylene microwavable container (control); PBS: PBS film (untreated); PBS + BM 1.5%: PBS film + 1.5% Biomaster silver; PBS + BM 3%: PBS film + 3% Biomaster silver; PBS + SAN 1%: PBS film + 1% SANAFOR; PBS + SAN 2%: PBS film + 2% SANAFOR. Error bars indicate standard deviation (n = 3). The different A–F capital letters are significantly different ($p < 0.05$) among treatment for each day. The different a–d lowercase letters are significantly different ($p < 0.05$) among storage day for each sample.

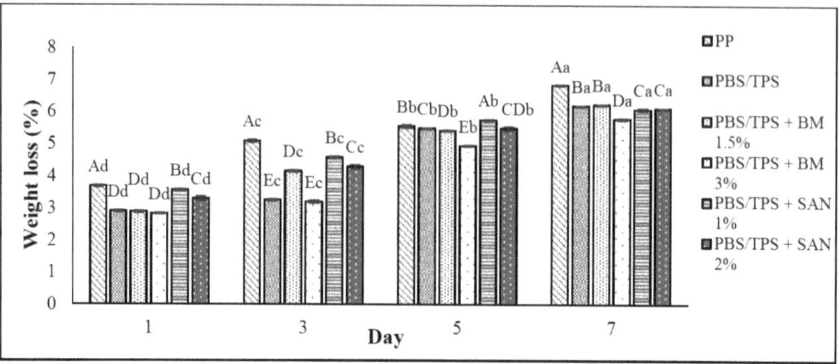

Figure 6. The weight loss values of chicken breast fillets stored in PP container and PBS/TPS film laminated tray during storage at 4 °C for seven days. Note: PP: polypropylene microwavable container (control); PBS/TPS: PBS/TPS film (untreated); PBS/TPS + BM 1.5%: PBS/TPS film + 1.5% Biomaster silver; PBS/TPS + BM 3%: PBS/TPS film + 3% Biomaster silver; PBS/TPS + SAN 1%: PBS/TPS film + 1% SANAFOR; PBS/TPS + SAN 2%: PBS/TPS film + 2% SANAFOR. Error bars indicate standard deviation (n = 3). The different A–E capital letters are significantly different ($p < 0.05$) among treatment for each day. The different a–d lowercase letters are significantly different ($p < 0.05$) among storage day for each sample.

3.4. Texture Analysis

The texture of the chicken breast fillet samples was analyzed in terms of hardness. The hardness values of the chicken breast fillet samples stored in the PBS and PBS/TPS film laminated tray during storage at 4 °C for seven days are shown in Figures 7 and 8, respectively. The hardness values increased ($p < 0.05$) throughout the seven days of the storage period for all samples, with values between 118.31 and 144.02% (Figure 7) and 116.22 and 144.02% (Figure 8). Compared to the control sample, the chicken samples stored in PBS and PBS/TPS film exhibited less hardness.

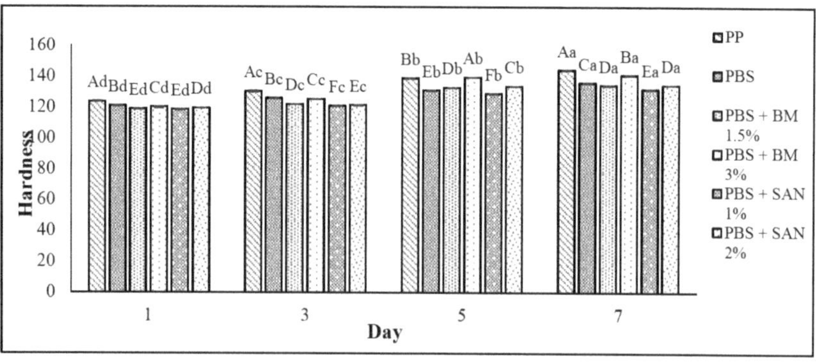

Figure 7. The hardness values of chicken breast fillets stored in PP container and PBS film laminated tray during storage at 4 °C for seven days. Note: PP: polypropylene microwavable container (control); PBS: PBS film (untreated); PBS + BM 1.5%: PBS film + 1.5% Biomaster silver; PBS + BM 3%: PBS film + 3% Biomaster silver; PBS + SAN 1%: PBS film + 1% SANAFOR; PBS + SAN 2%: PBS film + 2% SANAFOR. Error bars indicate standard deviation (n = 3). The different A–F capital letters are significantly different ($p < 0.05$) among treatment for each day. The different a–d lowercase letters are significantly different ($p < 0.05$) among storage day for each sample.

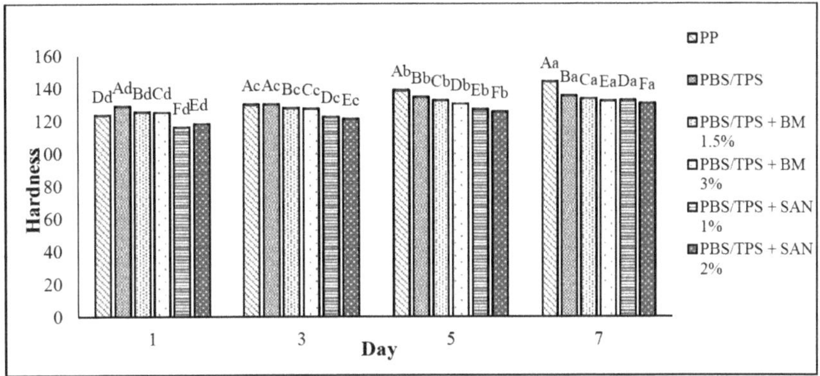

Figure 8. The hardness values of chicken breast fillets stored in PP container and PBS/TPS film laminated tray during storage at 4 °C for seven days. Note: PP: polypropylene microwavable container (control); PBS/TPS: PBS/TPS film (untreated); PBS/TPS + BM 1.5%: PBS/TPS film + 1.5% Biomaster silver; PBS/TPS + BM 3%: PBS/TPS film + 3% Biomaster silver; PBS/TPS + SAN 1%: PBS/TPS film + 1% SANAFOR; PBS/TPS + SAN 2%: PBS/TPS film + 2% SANAFOR. Error bars indicate standard deviation (n = 3). The different A–F capital letters are significantly different ($p < 0.05$) among treatment for each day. The different a–d lowercase letters are significantly different ($p < 0.05$) among storage day for each sample.

The lowest hardness of chicken fillets stored in the treated films might have been due to their lower weight loss, as discussed in Section 3.4, indicating that the treated films decreased the loss of weight, and then enhanced the texture of the chicken samples during storage at 4 °C. Based on Figure 7, the results show that the chicken samples stored in PBS + SAN 1% film recorded the lowest hardness value (131.48 ± 0.04) at the end of storage. Meanwhile, Figure 8 shows that the chicken samples stored in PBS/TPS + SAN 2% film had the lowest hardness value (130.57 ± 0.03).

3.5. Color Changes and Overall Visual Quality

In contemporary packing technologies, one of the primary assumptions is that the desired color would be preserved for as long as possible [40]. The overall visual quality of the chicken breast fillet samples stored in the PP container and PBS film laminated tray during storage at 4 °C for seven days is shown in Table 2. Table 2 shows that chicken samples stored in a tray laminated with PBS + SAN 1% film had the most pleasant appearance compared to the chicken samples stored in the PP container and other films. The chicken sample was still fresh on day 3 of storage and the color variation was not significant as compared to the other chicken samples, and no bad odor was detected once the lid was opened. The color (L*, a*, and b* values) of the chicken fillet samples kept in the PP container and PBS film laminated tray during storage at 4 °C for seven days is shown in Figure 9. In this study, the initial L*, a*, and b* values of the chicken fillets without any treatment (day 0) were 56.02 ± 0.21, 12.01 ± 0.08, and 10.29 ± 0.12, respectively.

According to Figure 9, the L* (lightness) and a* (redness) values for each chicken sample decreased ($p < 0.05$) gradually throughout the seven-day storage period storage. However, the chicken fillets presented increased ($p < 0.05$) b* (yellowness) values over the period of storage. The L* values were between 36.50 and 58.06, the a* values were between 7.00 and 11.33, and the b* values were between 10.33 and 15.29. The decrease in L* and a* values for all chicken fillets throughout the storage period of seven days indicates that the chicken darkened throughout the storage. The L* value of chicken samples stored in PBS film incorporated with BM and SAN was able to slow down the chicken darkening during the seven-day storage period compared to the PP container (control) and untreated PBS. The PBS + SAN 1% film exhibited the highest ($p < 0.05$) L* value, followed by PBS + SAN

2%, PBS + BM 3%, and PBS + BM 1.5% films. The PBS + SAN 1% also showed the highest ($p < 0.05$) a* value on day 7. The chicken samples stored in PBS + SAN 1% film recorded the lowest b* value ($p < 0.05$) compared to the chicken samples in other films.

The overall visual quality of the chicken breast fillet samples stored in the PP container and PBS/TPS film laminated tray during storage at 4 °C for seven days is shown in Table 3. Based on Table 3, the chicken samples stored in PBS/TPS + SAN 1% and PBS/TPS + SAN 2% had the most appealing appearance compared to the chicken samples in the PP container and other films. The chicken fillets were still fresh on the third day of storage, and the color variation did not significantly change. The color (L*, a*, and b* values) of the chicken fillet samples kept in the PP container and PBS/TPS film laminated tray during storage at 4 °C for seven days is shown in Figure 10.

According to Figure 10, the L* (lightness) and a* (redness) values for all chicken samples decreased ($p < 0.05$) gradually throughout the seven-days of the storage period. However, the chicken samples showed increased ($p < 0.05$) b* (yellowness) values over the storage period. The L* values were between 36.50 and 58.33, the a* values were between 6.95 and 11.90, and the b* values were between 10.33 and 15.29. The L* value of the chicken samples stored in PBS/TPS films incorporated with BM and SAN were able to slow down the darkening of the chicken samples throughout the seven days of the storage period compared to the PP container (control) and untreated PBS/TPS, with PBS/TPS + SAN 2% exhibiting the highest ($p < 0.05$) L* value, followed by PBS/TPS + SAN 1%, PBS/TPS + BM 3%, PBS/TPS + BM 1.5%, PBS/TPS, and PP. However, the chicken samples stored in PBS/TPS + BM3% film showed the highest ($p < 0.05$) a* value on day 7, followed by the PBS/TPS + BM 1.5% film. The chicken sample stored in the PBS/TPS + SAN 2% film showed the lowest ($p < 0.05$) b* value on day 7, followed by PBS/TPS + SAN 1%, PBS/TPS + BM 3%, PBS/TPS + BM 1.5%, PBS/TPS, and PP.

From the color analysis, it can be concluded that chicken breast fillets stored in PBS + SAN 1% and PBS/TPS + SAN 2% showed the least significant color variation throughout the seven days of the storage period compared to the chicken breast fillets stored in the PP container and other films. Both of these films are the most efficient for slowing the darkening behavior of chicken breast fillets resulting from the accumulation of ammonia and amines and retaining the red color of the chicken breast fillets [36].

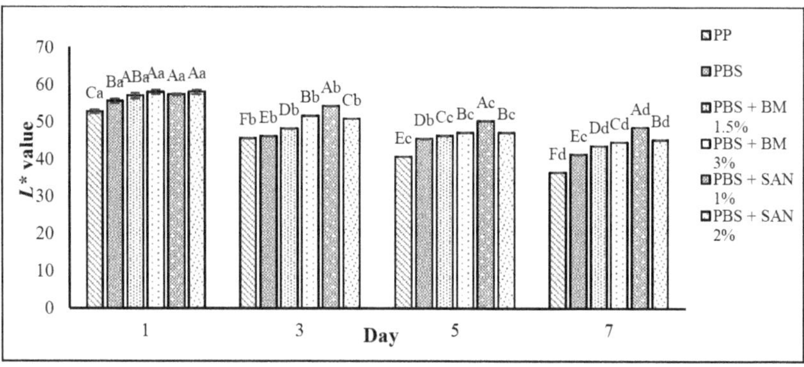

Figure 9. *Cont.*

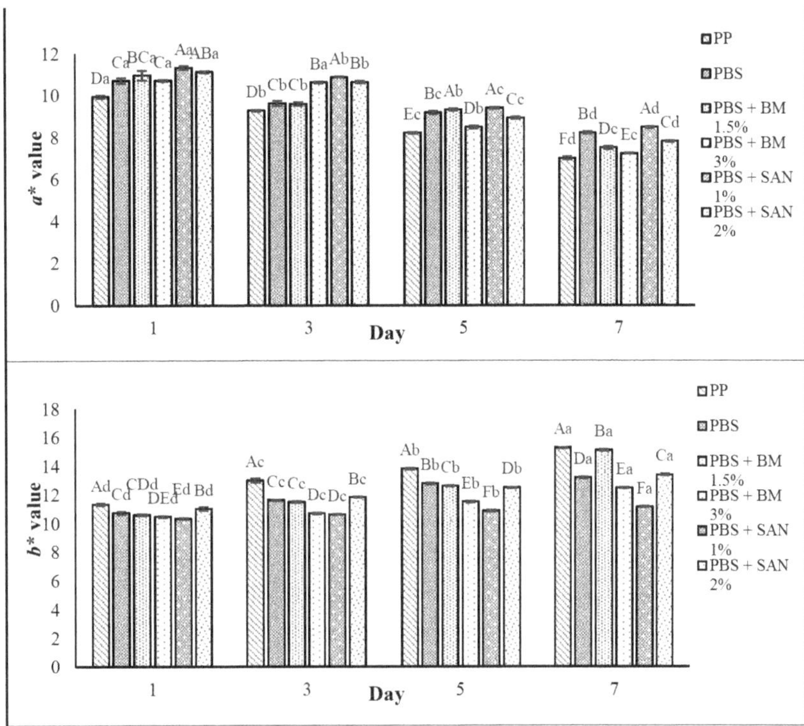

Figure 9. The L*, a*, and b* values of chicken breast fillets stored in PP container and PBS film laminated tray during storage at 4 °C for seven days. Note: Color parameters: L* (from 0 black to 100 white), a* (from -a* green to +a* red), and b* (from -b* blue to +b* yellow). PP: polypropylene microwavable container (control); PBS: PBS film (untreated); PBS + BM 1.5%: PBS film + 1.5% Biomaster silver; PBS + BM 3%: PBS film + 3% Biomaster silver; PBS + SAN 1%: PBS film + 1% SANAFOR; PBS + SAN 2%: PBS film + 2% SANAFOR. Error bars indicate standard deviation (n = 3). The different A–F capital letters are significantly different ($p < 0.05$) among treatment for each day. The different a–d lowercase letters are significantly different ($p < 0.05$) among storage day for each sample.

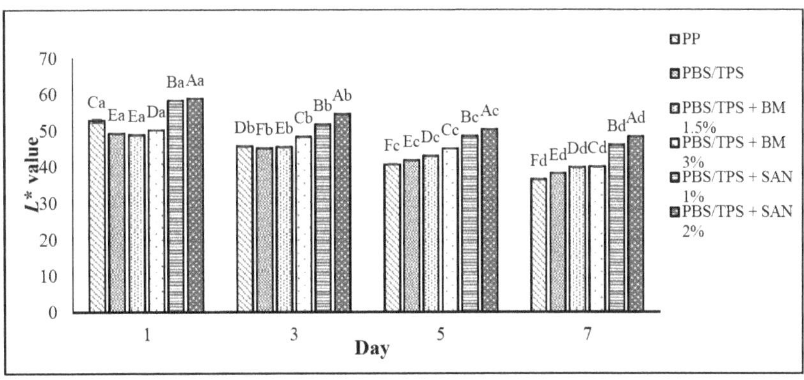

Figure 10. *Cont.*

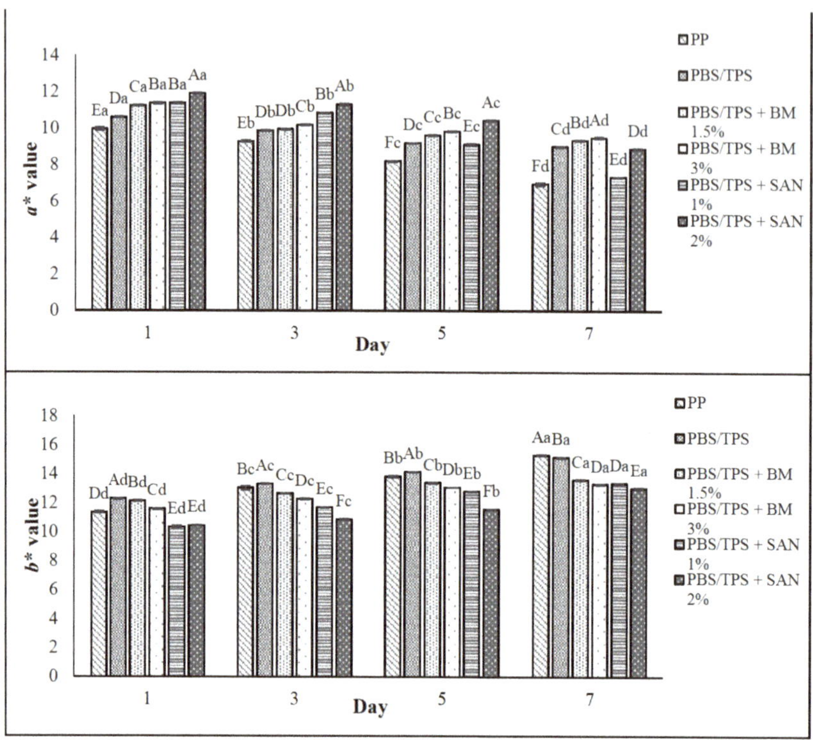

Figure 10. The L*, a*, and b* values of chicken breast fillets stored in PP container and PBS/TPS film laminated tray during storage at 4 °C for seven days. Note: Color parameters: L* (from 0 black to 100 white), a* (from -a* green to +a* red), and b* (from -b* blue to +b* yellow). PP: polypropylene microwavable container (control); PBS/TPS: PBS/TPS film (untreated); PBS/TPS + BM 1.5%: PBS/TPS film + 1.5% Biomaster silver; PBS/TPS + BM 3%: PBS/TPS film + 3% Biomaster silver; PBS/TPS + SAN 1%: PBS/TPS film + 1% SANAFOR; PBS/TPS + SAN 2%: PBS/TPS film + 2% SANAFOR. Error bars indicate standard deviation (n = 3). The different A–F capital letters are significantly different ($p < 0.05$) among treatment for each day. The different a–d lowercase letters are significantly different ($p < 0.05$) among storage day for each samples.

Table 2. The overall visual quality of chicken breast fillets stored in PP container and PBS film laminated tray during storage at 4 °C for seven days.

Day	0	1	3	5	7
PP	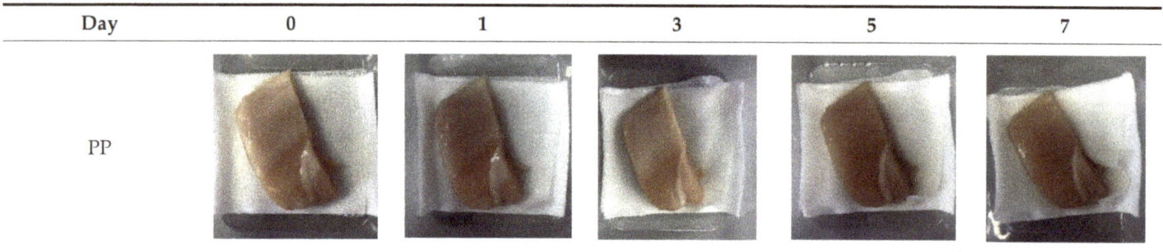				

Table 2. *Cont.*

Day	0	1	3	5	7
PBS	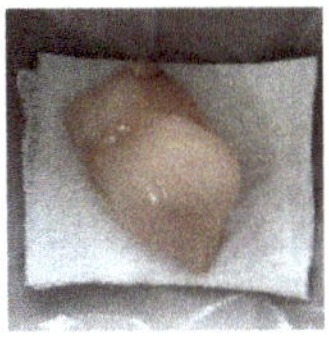		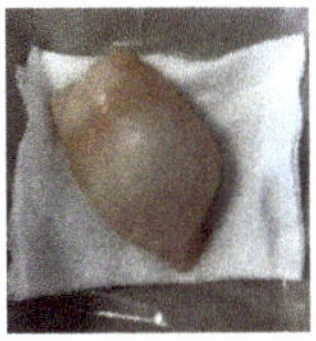	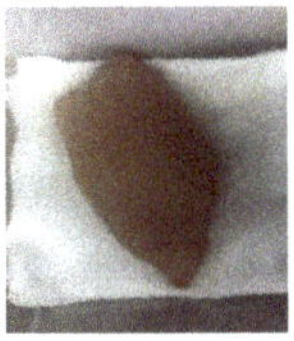	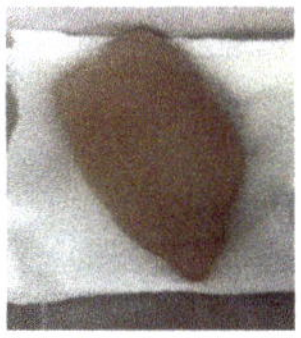
PBS + BM 1.5%	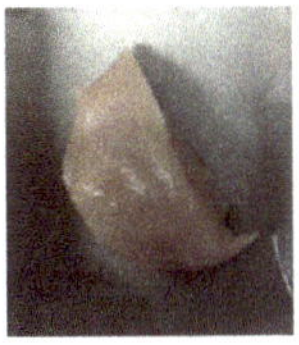	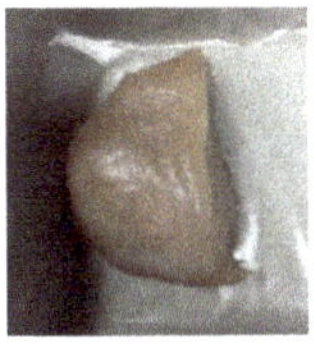	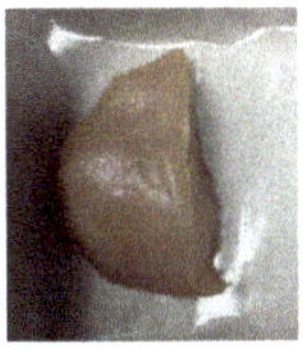	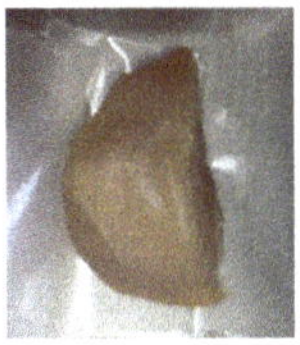	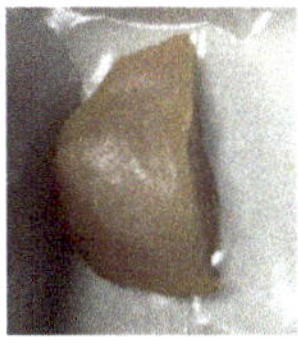
PBS + BM 3%	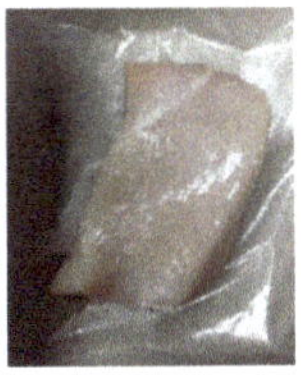	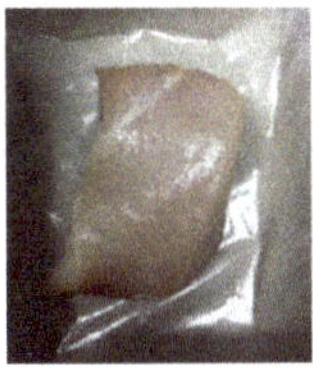		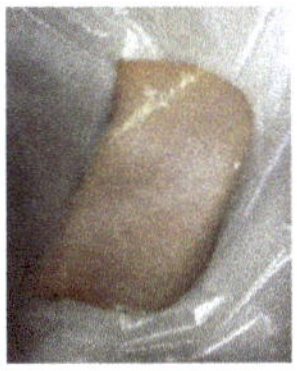	
PBS + SAN 1%	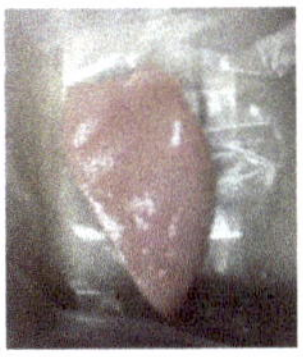	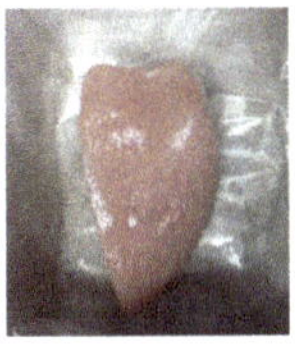	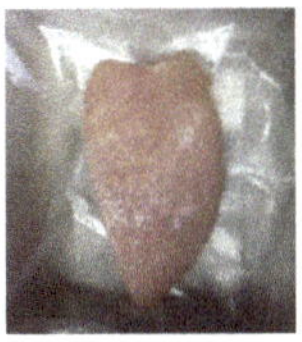	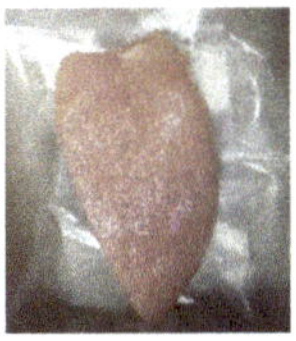	
PBS + SAN 2%	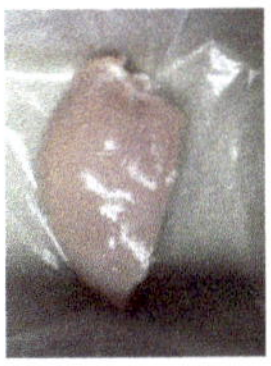	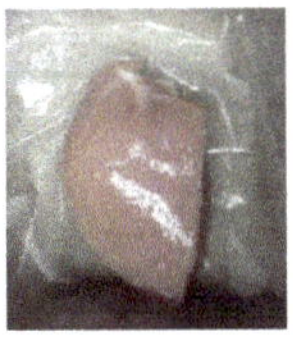	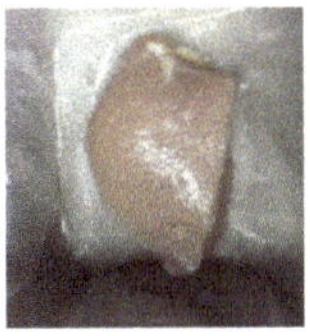	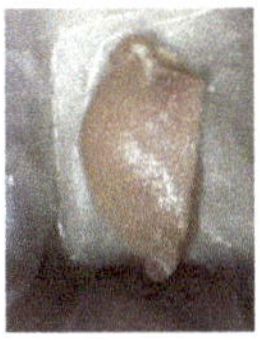	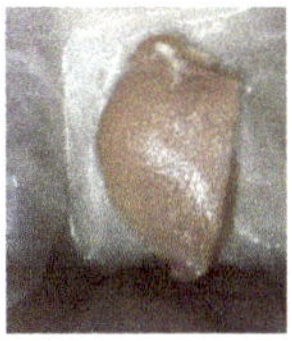

Note: PP: polypropylene microwavable container (control); PBS: PBS film (untreated); PBS + BM 1.5%: PBS film + 1.5% Biomaster silver; PBS + BM 3%: PBS film + 3% Biomaster silver; PBS + SAN 1%: PBS film + 1% SANAFOR; PBS + SAN 2%: PBS film + 2% SANAFOR.

Table 3. The overall visual quality of chicken breast fillets stored in PP container and PBS/TPS film laminated tray during storage at 4 °C for seven days.

Day	0	1	3	5	7
PP	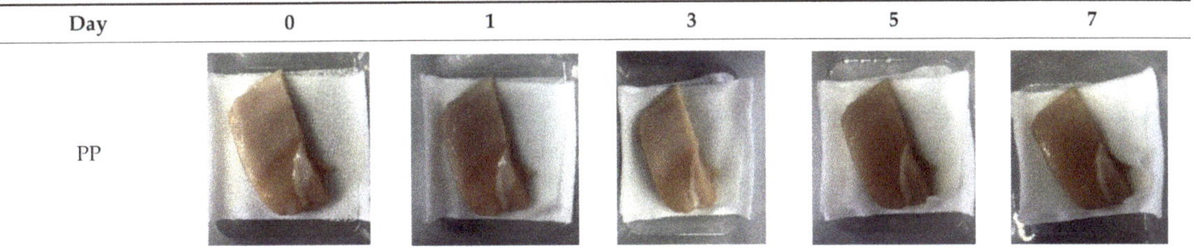				

Table 3. *Cont.*

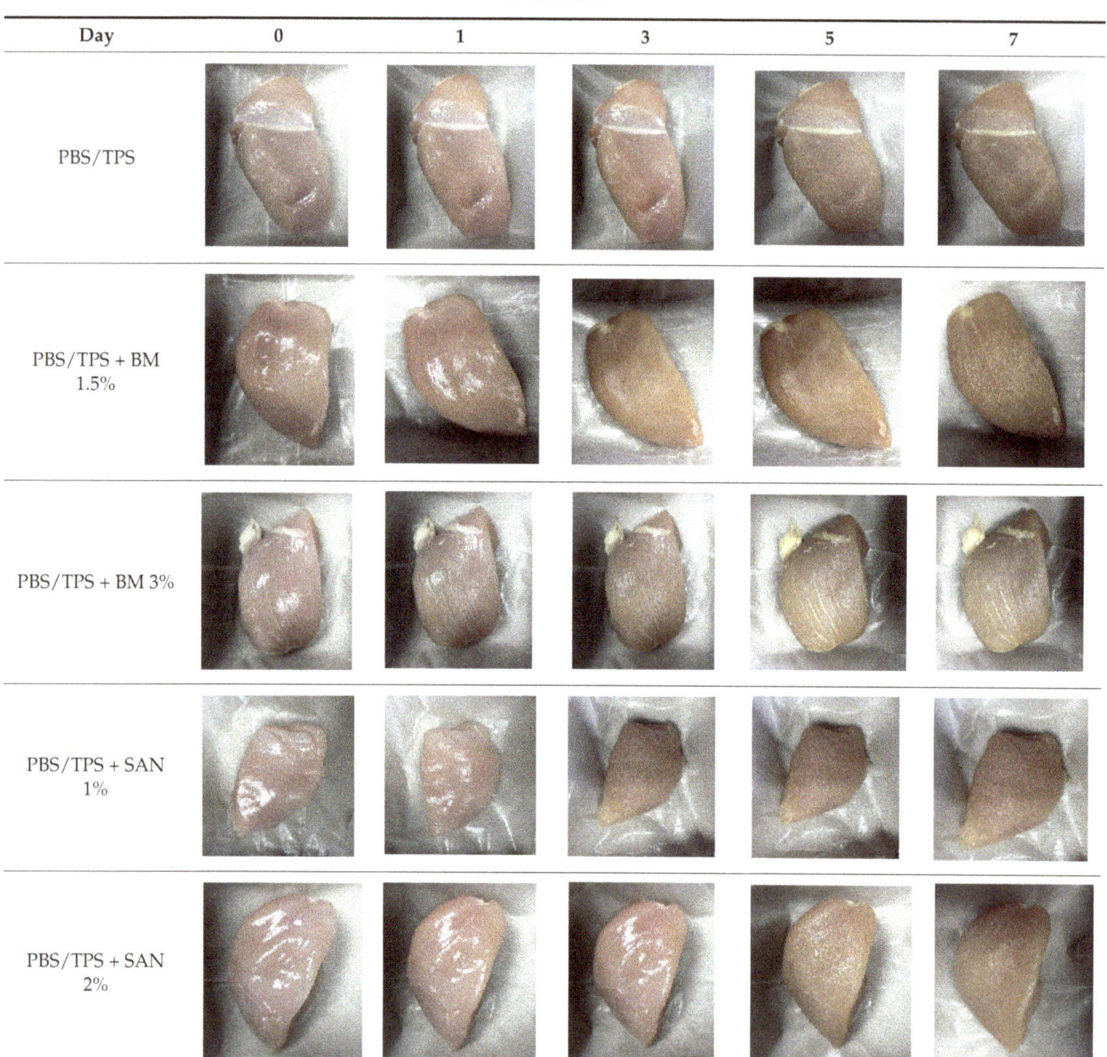

Day	0	1	3	5	7
PBS/TPS					
PBS/TPS + BM 1.5%					
PBS/TPS + BM 3%					
PBS/TPS + SAN 1%					
PBS/TPS + SAN 2%					

Note: PP: polypropylene microwavable container (control); PBS/TPS: PBS/TPS film (untreated); PBS/TPS + BM 1.5%: PBS/TPS film + 1.5% Biomaster silver; PBS/TPS + BM 3%: PBS/TPS film + 3% Biomaster silver; PBS/TPS + SAN 1%: PBS/TPS film + 1% SANAFOR; PBS/TPS + SAN 2%: PBS/TPS film + 2% SANAFOR.

4. Conclusions

In conclusion, the addition of TPS to PBS film showed a lower microbial log growth than PBS film alone. With antimicrobial agents BM and SAN being added to the packaging material, it further aids in monitoring the microbial growth and enhances the storability of chicken breast fillets up to seven days. However, the incorporation of 2% SAN into PBS film and 1% or 2% SAN into PBS/TPS films showed the highest potential as an antimicrobial property in reducing the log value compared to other film packaging. The quality assessment proved that the incorporation of BM and SAN into the films can enhance the quality of chicken breast fillets by minimizing the color variation, slowing pH increment, decreasing weight loss, and decelerating the hardening process of the chicken breast fillets throughout storage. The PBS + SAN 1% and PBS/TPS + SAN 2% films exhibited the least color variation. The PBS + SAN 1% and PBS/TPS + BM 3% films showed the least pH increment. The PBS + SAN 1%, PBS + BM 1.5%, and PBS/TPS + BM 3% films showed the least weight loss. The PBS + SAN 1% and PBS/TPS + SAN 2% films exhibited the least hardness. However, the most notable enhancement of the chicken breast fillets' quality was observed in the PBS film incorporated with 1% SAN, as it recorded the least color variation, pH increment, weight loss, and hardness. Therefore, it has potential to be used as new antimicrobial packaging material for chicken fillets and could be an alternative to plastic packaging.

Author Contributions: Conceptualization, N.L.Y., N.-A.A.M., H.F., A.A. and M.S.; methodology, N.A., U.K.N., A.H.N., H.F., L.K.K., M.S. and A.A.; validation, N.L.Y., N.-A.A.M. and, M.J.; formal analysis and investigation, N.A., U.K.N. and A.H.N.; writing—original draft preparation, N.L.Y., N.-A.A.M., N.A., U.K.N. and A.H.N.; writing—review and editing, N.A., H.F., M.J., A.A., M.S. and L.K.K.; supervision and project administration, N.L.Y., N.-A.A.M. and M.J.; funding acquisition, M.J. and H.F. All authors have read and agreed to the published version of the manuscript.

Funding: This project was supported by the Newton-Ungku Omar Fund (Project Vot No.: 6300873). This work was funded by Researchers Supporting Project number (RSP-2021/117), King Saud University, Riyadh, Saudi Arabia.

Institutional Review Board Statement: Not applicable.

Informed Consent Statement: Not applicable.

Data Availability Statement: Not applicable.

Acknowledgments: We are thankful to the Department of Food Science and Technology, Universiti Putra Malaysia, for providing us with facilities to carry out the experimental work. Authors are also thankful to the Researchers Supporting Project number (RSP-2021/117), King Saud University, Riyadh, Saudi Arabia for funding this work.

Conflicts of Interest: The authors declare no conflict of interest.

References

1. Malhotra, B.; Keshwani, A.; Kharkwal, H. Antimicrobial food packaging: Potential and pitfalls. *Front. Microbiol.* **2015**, *6*, 611. [CrossRef]
2. Quintavalla, S.; Vicini, L. Antimicrobial food packaging in meat industry. *Meat Sci.* **2002**, *62*, 373–380. [CrossRef]
3. Vaithiyanathan, S.; Naveena, B.M.; Muthukumar, M.; Girish, P.S.; Kondaiah, N. Effect of dipping in pomegranate (*Punica granatum*) fruit juice phenolic solution on the shelf life of chicken meat under refrigerated storage (4 °C). *Meat Sci.* **2011**, *88*, 409–414. [CrossRef]
4. Konuk, D.T.; Korel, F. Active packaging films as a carrier of black cumin essential oil: Development and effect on quality and shelf-life of chicken breast meat. *Food Packag. Shelf Life* **2019**, *19*, 210–217. [CrossRef]
5. Díaz-Galindo, E.P.; Nesic, A.; Cabrera-Barjas, G.; Mardones, C.; von Baer, D.; Bautista-Baños, S.; Dublan Garcia, O. Physical-chemical evaluation of active food packaging material based on thermoplastic starch loaded with grape cane extract. *Molecules* **2020**, *25*, 1306. [CrossRef]
6. Liu, W.G.; Zhang, X.C.; Li, H.Y.; Liu, Z. Effect of surface modification with 3-aminopropyltriethyloxy silane on mechanical and crystallization performances of ZnO/poly(butylenesuccinate) composites. *Compos. Part B Eng.* **2012**, *43*, 2209–2216. [CrossRef]
7. Pandey, J.K.; Reddy, K.R.; Kumar, A.P.; Singh, R.P. An overview on the degradability of polymer nanocomposites. *Polym. Degrad. Stab.* **2005**, *88*, 234–250. [CrossRef]
8. Nam, T.H.; Ogihara, S.; Tung, N.H.; Kobayashi, S. Effect of alkali treatment on interfacial and mechanical properties of coir fiber reinforced poly(butylene succinate) biodegradable composites. *Compos. Part B Eng.* **2011**, *42*, 1648–1656. [CrossRef]
9. Khan, B.; Niazi, M.B.K.; Samin, G.; Jahan, Z. Thermoplastic Starch: A Possible Biodegradable Food Packaging Material—A Review. *J. Food Process Eng.* **2017**, *40*, e12447. [CrossRef]
10. Fahrngruber, B.; Fortea-Verdejo, M.; Wimmer, R.; Mundigler, N. Starch/Poly(butylene succinate) compatibilizers: Effect of different reaction-approaches on the properties of thermoplastic starch-based compostable films. *J. Polym. Environ.* **2020**, *28*, 257–270. [CrossRef]
11. Yin, Q.J.; Chen, F.P.; Zhang, H.; Liu, C.S. Mechanical properties and thermal behavior of TPS/PBS Blends with maleated PBS as a compatibilizer. *Adv. Mater. Res.* **2015**, *1119*, 306–309. [CrossRef]

12. Petchwattana, N.; Covavisaruch, S.; Wibooranawong, S.; Naknaen, P. Antimicrobial food packaging prepared from poly(butylene succinate) and zinc oxide. *Measurement* **2016**, *93*, 442–448. [CrossRef]
13. Llorens, A.; Lloret, E.; Picouet, P.A.; Trbojevich, R.; Fernandez, A. Metallic-based micro and nanocomposites in food contact materials and active food packaging. *Trends Food Sci. Technol.* **2012**, *24*, 19–29. [CrossRef]
14. Zhao, S.; Yao, J.; Fei, X.; Shao, Z.; Chen, X. An antimicrobial film by embedding in situ synthesized silver nanoparticles in soy protein isolate. *Mater. Lett.* **2013**, *95*, 142–144. [CrossRef]
15. Theivendran, S.; Hettiarachchy, N.S.; Johnson, M.G. Inhibition of Listeria monocytogenes by nisin combined with grape seed extract or green tea extract in soy protein film coated on Turkey Frankfurters. *J. Food Sci.* **2006**, *71*, 39–44. [CrossRef]
16. Mauriello, G.; De Luca, E.; La Storia, A.; Villani, F.; Ercolini, D. Antimicrobial activity of a nisin-activated plastic film for food packaging. *Lett. Appl. Microbiol.* **2005**, *41*, 464–469. [CrossRef] [PubMed]
17. Sivarooban, T.; Hettiarachchy, N.S.; Johnson, M.G. Physical and antimicrobial properties of grape seed extract, nisin, and EDTA incorporated soy protein edible films. *Food Res. Int.* **2008**, *41*, 781–785. [CrossRef]
18. Calatayud, M.; López-de-Dicastillo, C.; López-Carballo, G.; Vélez, D.; Hernández Muñoz, P.; Gavara, R. Active films based on cocoa extract with antioxidant, antimicrobial and biological applications. *Food Chem.* **2013**, *139*, 51–58. [CrossRef] [PubMed]
19. Rojas-Graü, M.A.; Avena-Bustillos, R.J.; Olsen, C.; Friedman, M.; Henika, P.R.; Martín-Belloso, O.; Pan, Z.; McHugh, T.H. Effects of plant essential oils and oil compounds on mechanical, barrier and antimicrobial properties of alginate–apple puree edible films. *J. Food Eng.* **2007**, *81*, 634–641. [CrossRef]
20. Holt, K.B.; Bard, A.J. Interaction of silver(i) ions with the respiratory chain of Escherichia coli: An electrochemical and scanning electrochemical microscopy study of the antimicrobial mechanism of micromolar Ag+. *Biochemistry* **2005**, *44*, 13214–13223. [CrossRef] [PubMed]
21. Xu, H.; Qu, F.; Xu, H.; Lai, W.; Wang, Y.A.; Aguilar, Z.P.; Wei, H. Role of reactive oxygen species in the antibacterial mechanism of silver nanoparticles on *Escherichia coli* O157: H7. *Biometals* **2012**, *25*, 45–53. [CrossRef]
22. Matés, J.M.; Sánchez-Jiménez, F.M. Role of reactive oxygen species in apoptosis: Implications for cancer therapy. *Int. J. Biochem. Cell Biol.* **2000**, *32*, 157–170. [CrossRef]
23. Masoumi, B.; Abbasi, A.; Mazloomi, S.M. The effect of saffron on microbial, physicochemical and texture profile of chicken (breast) meat stored in refrigerator. *Int. J. Nutr. Sci.* **2018**, *3*, 164–170.
24. Wang, J.; Zhuang, H.; Hinton, A.; Zhang, J. Influence of in-package cold plasma treatment on microbiological shelf life and appearance of fresh chicken breast fillets. *Food Microbiol.* **2016**, *60*, 142–146. [CrossRef]
25. Mohammadi, H.; Kamkar, A.; Misaghi, A.; Zunabovic-Pichler, M.; Fatehi, S. Nanocomposite films with CMC, okra mucilage, and ZnO nanoparticles: Extending the shelf-life of chicken breast meat. *Food Packag. Shelf Life* **2019**, *21*, 100330. [CrossRef]
26. Senter, S.D.; Arnold, J.W.; Chew, V. APC values and volatile compounds formed in commercially processed, raw chicken parts during storage at 4 and 13 °C and under simulated temperature abuse conditions. *J. Sci. Food Agric.* **2000**, *80*, 1559–1564. [CrossRef]
27. Bian, J.; Han, L.; Wang, X.; Wen, X.; Han, C.; Wang, S.; Dong, L. Nonisothermal crystallization behavior and mechanical properties of poly(butylene succinate)/silica nanocomposites. *J. Appl. Polym. Sci.* **2010**, *116*, 902–912. [CrossRef]
28. Appendini, P.; Hotchkiss, J.H. Review of antimicrobial food packaging. *Innov. Food Sci. Emerg. Technol.* **2002**, *3*, 113–126. [CrossRef]
29. Warsiki, E.; Bawardi, J.T. Assessing mechanical properties and antimicrobial activity of zinc oxide-starch biofilm. *IOP Confer. Ser. Earth Environ. Sci.* **2018**, *209*, 012003. [CrossRef]
30. Cardoso, L.G.; Pereira Santos, J.C.; Camilloto, G.P.; Miranda, A.L.; Druzian, J.I.; Guimarães, A.G. Development of active films poly (butylene adipate co-terephthalate)—PBAT incorporated with oregano essential oil and application in fish fillet preservation. *Ind. Crop. Prod.* **2017**, *108*, 388–397. [CrossRef]
31. Ahmed, J.; Mulla, M.Z.; Arfat, Y.A. Thermo-mechanical, structural characterization and antibacterial performance of solvent casted polylactide/cinnamon oil composite films. *Food Control* **2016**, *69*, 196–204. [CrossRef]
32. Shahbazi, Y.; Shavisi, N. A novel active food packaging film for shelf-life extension of minced beef meat. *J. Food Saf.* **2018**, *38*, 12569. [CrossRef]
33. Emma, G.; Lesley, T. Using silver to help combat Campylobacter and other bacteria. *Perspect. Public Health* **2013**, *133*, 292–293.
34. Kumar, R.; Ghoshal, G.; Goyal, M. Biodegradable composite films/coatings of modified corn starch/gelatin for shelf life improvement of cucumber. *J. Food Sci. Technol.* **2020**, *58*, 1227–1237. [CrossRef] [PubMed]
35. Rashidaie, S.S.A.; Ariaii, P.; Charmchian Langerodi, M. Effects of encapsulated rosemary extract on oxidative and microbiological stability of beef meat during refrigerated storage. *Food Sci. Nutr.* **2019**, *7*, 3969–3978. [CrossRef] [PubMed]
36. Katiyo, W.; de Kock, H.L.; Coorey, R.; Buys, E.M. Sensory implications of chicken meat spoilage in relation to microbial and physicochemical characteristics during refrigerated storage. *LWT* **2020**, *128*, 109468. [CrossRef]
37. Ashrafi, A.; Jokar, M.; Mohammadi Nafchi, A. Preparation and characterization of biocomposite film based on chitosan and kombucha tea as active food packaging. *Int. J. Biol. Macromol.* **2018**, *108*, 444–454. [CrossRef]
38. Souza, V.G.L.; Pires, J.R.A.; Vieira, É.T.; Coelhoso, I.M.; Duarte, M.P.; Fernando, A.L. Shelf life assessment of fresh poultry meat packaged in novel bionanocomposite of chitosan/montmorillonite incorporated with ginger essential oil. *Coatings* **2018**, *8*, 177. [CrossRef]

39. Amjadi, S.; Emaminia, S.; Nazari, M.; Davudian, S.H.; Roufegarinejad, L.; Hamishehkar, H. Application of Reinforced ZnO Nanoparticle-Incorporated Gelatin Bionanocomposite Film with Chitosan Nanofiber for Packaging of Chicken Fillet and Cheese as Food Models. *Food Bioprocess Technol.* **2019**, *12*, 1205–1219. [CrossRef]
40. Gazalli, H.; Malik, A.H.; Jalal, H.; Afshan, S.; Mir, A.; Ashraf, H. Packaging of meat. *Int. J. Food Nutr. Saf.* **2013**, *4*, 70–80.

Article

Effect of Liquid Absorbent Pads and Packaging Parameters on Drip Loss and Quality of Chicken Breast Fillets

Marit Kvalvåg Pettersen [1,*], Julie Nilsen-Nygaard [1], Anlaug Ådland Hansen [1], Mats Carlehög [1] and Kristian Hovde Liland [2]

[1] Norwegian Institute of Food, Fisheries and Aquaculture (Nofima AS), Osloveien 1, 1430 Ås, Norway; post@nofima.no (J.N.-N.); Anlaug.Hansen@Nofima.no (A.Å.H.); Mats.Carlehog@Nofima.no (M.C.)
[2] Faculty of Science and Technology, Norwegian University of Life Sciences (NMBU), 1432 Ås, Norway; Kristian.Liland@nmbu.no
* Correspondence: marit.kvalvag.pettersen@nofima.no; Tel.: +47-64970280 or +47-92807951

Abstract: Visible liquid inside food packages is perceived as unattractive to consumers, and may result in food waste—a significant factor that can compromise sustainability in food value chains. However, an absorber with overdimensioned capacity may cause alterations in texture and a dryer product, which in turn may affect consumers' satisfaction and repurchase. In this study we compared the effect of a number of liquid absorbent pads in combination with headspace gas composition (60% CO_2/40% N_2 and 75% O_2/25% CO_2) and gas-to-product volume ratio (g/p) on drip loss and quality of fresh chicken breast fillets. A significant increase in drip loss with an increasing number of liquid absorbent pads was documented. The increase was more pronounced in 60% CO_2/40% N_2 compared to 75% O_2/25% CO_2. By comparing packaging variants with a different number of liquid absorbent pads, a higher drip loss for all tested was found at g/p 1.8 compared to g/p 2.9. Total viable counts (TVC) were independent of whether there was free liquid in contact with the product, and TVC was independent of gas composition. Differentiation between the gas compositions was seen for specific bacterial analyses. While significant changes were observed using texture analysis, sensory evaluation of the chicken breast fillets did not show any negative effect in texture related attributes. This study demonstrates the importance of optimized control of meat drip loss, as product-adjusted liquid absorption may affect economy, food quality, and consumer satisfaction, as well as food waste.

Keywords: food packaging; drip loss; liquid absorbent pad; chicken breast fillet; texture; sensory evaluation

Citation: Pettersen, M.K.; Nilsen-Nygaard, J.; Hansen, A.Å.; Carlehög, M.; Liland, K.H. Effect of Liquid Absorbent Pads and Packaging Parameters on Drip Loss and Quality of Chicken Breast Fillets. *Foods* **2021**, *10*, 1340. https://doi.org/10.3390/foods10061340

Academic Editors: Valeria Rizzo and Muratore Giuseppe

Received: 3 May 2021
Accepted: 8 June 2021
Published: 10 June 2021

1. Introduction

In light of the past several years' focus on sustainability, and the global targets of the United Nations Sustainable Development goals [1] related to responsible production and consumption, the drive towards development of food systems for reducing food loss and food waste has been pronounced. Packaging in general, and especially plastics, have for the last years seen increasing public awareness of the related environmental challenges, specifically related to littering and marine debris [2]. However, one of the main functions of packaging is to protect and preserve the food in the total value chain—from the producer to the consumer. Food packaging is recognized to contribute to food waste reduction and more sustainable food value chains [1,3–5].

Through evaluation of the environmental impact of meat products, measured as greenhouse gas (GHG) emissions [6,7], it has been found that the packaging is only responsible for a small part of the GHG emissions [8]. Considering the small environmental footprint of the packaging compared to that of the meat products, it is clear that optimal packaging systems to avoid food loss in the value chain and food waste at consumers should be high priority, and will contribute to more sustainable food systems.

Fresh meat products contain high amounts of water. The physiological water content in muscle foods, such as chicken, beef, and pork, is approximately 75% [9]. The water-holding capacity (WHC) of the meat refers to the ability of the meat to retain its natural or added water content during postmortem processing and storage [10,11]. The high water content of meat makes these products particularly prone to microbial spoilage, making them highly perishable. Another important factor tightly linked to the high water content, and thereby also to the perceived quality of these products, is their drip loss [12]. Unavoidably, these products will exude some liquid during storage. Scientific literature in the field mainly focuses on how different aspects of prehandling can be related to excessive WHC and drip loss for red meat (pork, beef, and lamb) [13,14]. Factors like genotype, feeding, slaughtering, and chilling have been evaluated [14], as well as the effect of cutting of muscle fibers. The only packaging-related aspects studied in relation to drip loss were the effect of shrinking or non-shrinking films [13], i.e., there are relatively few studies focusing on the effect of packaging variables [12,15,16]. However, in all cases the comparison has been between MAP (80% O_2/20% CO_2 or 30% CO_2/70% N_2) and vacuum [12,16]. Payne et al. (1998) compared vacuum with CO_2 flushing and/or the use of active packaging as an oxygen scavenger [15]. The studied storage temperatures have also varied a great deal, but no studies included storage at traditional/recommended storage temperatures at retailers in Europe [12,15,16]. Thus, to our knowledge, the effects of different MA gas compositions or gas-to-product volume ratios in relation to drip loss have not been reported for chicken meat.

A liquid absorber is often used in packages with fresh meat and fish to improve the appearance of the product. The capacity of these liquid absorbers is typically chosen by the food producers to ensure absorption of all drip loss, and may not be specifically designed for each product. An absorber with an overdimensioned capacity may cause an unnecessarily high drip loss. This can cause a dryer product and an alteration in texture. Furthermore, visible liquid in the packages can be perceived as unattractive by the consumer [17]. Sensory quality attributes such as juiciness and tenderness of the meat may be reduced, and these are important in terms of how the product is perceived by the consumer at the time of consumption [17].

There is an economic aspect to striving to limit the drip loss of muscle foods. Firstly, free liquid inside packages may be perceived as unattractive by the consumer, and may result in reduced sales [17,18]. In addition to this, a common perception seems to be that excess liquid inside packages can give rise to increased microbial growth and reduced meat quality. However, this has been disproven in a previous publication, where no such relation could be documented [18]. On the contrary, the study showed that the most attractive growth medium for bacteria is the product itself, not its exudate. Finally, the liquid lost implies a reduction in product weight and a reduced product yield for food producers.

Another aspect in the context of drip loss is the fact that the EU regulates the use of absorbers containing superabsorbent polymers (SAP), which are considered to be active packaging devices. For non-sealed absorbers it is mandatory for food producers to use absorbers of adequate capacity in order to ensure that the absorber can absorb all liquid lost from the product. This is to ensure that there is no leakage of SAP that can come into contact with the product (Commission Regulation (EC) No 450/2009 on active and intelligent materials and articles intended to come into contact with food) [19]. Due to this, food producers may be prone to choose an absorber that has an overdimensioned capacity for the product, in order to make sure that there is no free liquid inside the packages.

A high amount of CO_2 inside modified atmosphere packaging (MAP) is often associated with increased drip loss of the product, as CO_2 dissolving into the product causes the WHC to decrease [20–22]. One of the assumed mechanisms at play is the reduction of pH in the presence of CO_2 [23]. Others have reported negative correlation between CO_2 content and drip loss of meat [24]. However, often overlooked is the more pronounced effect of underpressure formation inside the packages at high CO_2 levels. CO_2 absorbed by the product may cause package deformation and a physical squeeze on the product, resulting in increased drip loss [25,26].

In this study we wanted to systematically investigate which packaging parameters are the most determinant for influencing the drip loss—and consequently, the physicochemical, microbiological, and sensory quality—of chicken breast fillets. The initial experimental setup was designed to address the following research questions: How does the number of liquid absorbent pads affect the drip loss and quality of the meat? How is the drip loss affected by different gas atmosphere compositions in MAP? Chicken breast fillets were chosen as the model product due to their relatively high drip loss and well-known challenges related to rapid microbiological spoilage. The number of liquid absorbent pads and different gas compositions were the variables included in the main experiment. Based on the results from the first experimental setup, a follow-up experiment was designed, aimed at addressing the new emerging research question: How is the drip loss affected in MAP with CO_2 at different gas-to-product volume (g/p) ratios? In this experiment the number of liquid absorbent pads and the gas-to-product (g/p) volume ratio were the main variables.

2. Materials and Methods

2.1. Sample Preparation and Storage Conditions

Chicken breast fillets (breast fillet tenderloin; pectoralis major) were obtained from a local producer using a fast and highly automated process with low temperature during the process. The slaughtering was performed in the morning with continuous cooling and cutting. The approximate time for slaughtering and cooling was 3 h, followed by cutting and packaging within less than 30 min. The fillets were wrapped in plastic bags and transported chilled in distribution boxes containing approximately 10 kg each. The packaging was performed at the research institute shortly after reception and within 48 h after slaughtering. The fillets were randomly selected from the distribution boxes. Two fillets with a total average weight of 339 g (339.1 $\pm$ 4.4 g) (329.9–350.9 g) were packaged in each tray. The samples were stored in dark conditions at 4 °C.

2.2. Packaging Materials

The chicken breast fillets were packaged in thermoformed trays with a base web consisting of amorphous polyethylene terephthalate/polyethylene (APET/PE) (Multipet 550 µm, Wipak, Nastola, Finland). Biaxer 65 XX HFP AFM consisting of polyethylene terephthalate/polyethylene/ethylene vinyl alcohol/polyethylene (PET/PE/EVOH/PE) (Wipak (Nastola, Finland)) was applied as the top web.

The oxygen transmission rates (OTR) of the materials were, according to the producer: 10 $cm^3/(m^2\ d)$ at 23 °C, 50% RH for the base web, and 5 $cm^3/(m^2\ d)$ at 23 °C, 50% RH for the top web.

The trays were thermoformed using a Multivac R145 thermoforming machine (Multivac, Wolfertschwenden, Germany).

In both experiments, different numbers of liquid absorbent pads were used—0, 1, and 2 (Absorber type 109642, MP-2501 75 × 115 mm black, Færch, Denmark)—and thereby 3 different possibilities of liquid absorption for each gas composition.

2.3. Packaging Methods and Experimental Design

The studies encompass two experiments with chicken breast fillets. In the first and main experiment (all analyses included) (hereafter referred to as Experiment 1), the gas composition and the number for liquid absorbent pads were the experimental design factors. In this experiment, the chicken breast fillets were stored in a modified atmosphere of 60% CO_2/40% N_2 or 25% CO_2/75% O_2, and a gas-to-product volume (g/p) ratio of 1.8 was applied for all samples. Four replicates of each sample type were prepared for each sampling time, performed after 0, 6, 14, and 20 days of storage.

In the second experiment (Experiment 2), the effect of the g/p ratio and the number of liquid absorbent pads was investigated. The chicken breast fillets were stored in 60% CO_2/40% N_2, and two different tray sizes were used. Trays with a volume of 860 mL resulted in an initial gas-to product-volume ratio (g/p ratio) of approximately 1.8, while trays with a volume of 1390 mL gave an initial g/p ratio of approximately 2.9. Four replicates of each sample type were prepared for each sampling time, performed after 0, 7, 14, and 20 days of storage.

2.4. *Analyses*

2.4.1. Headspace Gas Analyses and Drip Loss

The headspace atmosphere of the MA packages was analyzed for CO_2 and O_2 levels (%) immediately after packaging and at each sampling time using a CheckMate 9900 O_2/CO_2 analyzer (PBI Dansensor, Ringsted, Denmark). Gas was removed from the packaging for analysis using a needle through self-sealing patches on the packages.

Drip loss was determined by initially weighing the meat, the package, and the absorbent pad(s), and calculating the increase in weight of the packages (including the absorbent pads) at each sampling. Results are given as the percentage (%) of initial muscle weight, and refer to the corresponding drip loss from the meat. These analyses were performed in all experiments.

2.4.2. Texture Analyses and Dry Matter Content

Warner–Bratzler shear force (WBSF) and dry matter content measurements were performed (in Experiment 1) for chicken breast fillets cooked after storage in different packaging conditions after 0, 6, 14, and 21 days of storage. The fillets were vacuum packed in PA/PE (70 μm) (Maskegruppen, Norway) bags and heat treated in a water bath at 70 °C for 50 min before being cooled in ice water for 50 min. The samples were stored in the vacuum bags at 4 °C until the next day. Prior to WBSF measurements, the temperature of the samples was equilibrated at 20 °C for 1 hour. The fillets were cut into rectangular pieces of 1 × 1 × 2 cm along the fiber direction. The samples were sheared perpendicularly to the fiber direction with a triangular device attached to an Instron Materials Testing Machine (model 4202, Instron Engineering Co., High Wycombe, UK). The average maximum force (given as N/cm^2) was obtained from measurement of 6 replicates.

For determination of dry matter content, the samples were macerated/homogenized, and approximately 6 ± 0.5 g of the mass was accurately weighed into Petri dishes and oven dried at 105 °C for 18 h. The samples were weighed after drying, and the weight loss during drying was equal to the water content of the samples. The dry matter content was calculated as the percentage of the initial weight minus the water content (Dry matter content (%) = w_{sample} (%) − w_{water} (%)). Two replicates per sample variant were measured at each sampling time.

2.4.3. Microbiological Analyses

The selected microbiological analyses for chicken breast fillets (Experiments 1 and 2) were total viable count (TVC), *Enterobacteriaceae*, lactic acid bacteria (LAB), and *Brochothrix thermosphacta*, performed at the time of packaging and after the selected sampling time.

Samples of 3 × 3 cm^2 and 1 cm depth were cut with a sterile scalpel from the surface of the meat, weighed, macerated, and diluted 1:10 with peptone water and spread using a Whitley Automated Spiral Plater (WASP) (Don Whitley Scientific Ltd., West Yorkshire, UK). In addition, in Experiment 1, after 14 days of storage, samples from the bottom surfaces of the fillets were cut and included for analyses. Appropriate 10-fold dilutions were spread in duplicate on PCA (plate count agar; Difco, Difco Laboratories, Detroit, MI, USA) for total viable counts (TVC) (incubation temperature 30 °C, 72 h, anaerobic incubation), and on MRS agar (Man, Sharpe and Rogosa agar, MRS; Oxoid, Unipath Ltd., Basingstoke, Hampshire, UK) for lactic acid bacteria (incubated at 20 °C, 48 h, anaerobic incubation). *Enterobacteriaceae* were analyzed by use of VRBGA (Violet Red Bile Glucose Agar, Oxoid,

Hampshire, UK) (37 °C, 24 h, semi-aerobic conditions, cells embedded in agar with sterile overlay). *Brochothrix thermosphacta* was detected by use of STAA agar (streptomycin thallous acetate actidione) and an agar base (CM 0881 with selective supplement SR 0151E, Oxoid, Hampshire, UK) (25 °C for 48 h, aerobic incubation). Microbial counts are expressed as colony-forming units (cfu) per g.

2.4.4. Sensory Analysis

Sensory analysis was performed on both raw and heat-treated chicken breast fillets after 14 days of storage in both modified atmospheres—25% CO_2/75% O_2, and 60% CO_2/40% O_2—with 0, 1, and 2 liquid absorbent pads (g/p 1.8) (Experiment 1). A highly trained panel of 10 assessors (10 women; aged, 37–64 years) at Nofima (Ås, Norway) performed a sensory descriptive analysis (DA) according to the "generic descriptive analysis" [27] and ISO standard 13299 [28]. The assessors are regularly tested and trained according to ISO standard 8586, and the sensory laboratory follows the practice of ISO standard 8589 [29,30].

In a pretest session before the main test, the assessors were calibrated on samples that were considered the most different on the selected attributes typical for raw and heat-treated chicken fillets. The results from the pretest were evaluated and discussed by the panel leader and the assessors. This calibration procedure was performed in order to arrive at a common understanding and agreement of the selected attributes. This is common practice, with the purpose being to ensure that the assessors have a common understanding of how to evaluate and rank the different sensory attributes, and to obtain consensus for each attribute among the assessors. For raw evaluation the assessors agreed upon six sensory attributes describing odor: sourness odor, metallic odor, cloying odor, sulfurous odor, fermented odor, and chlorine odor. For heat-treated evaluation, the assessors agreed upon eight sensory attributes describing flavor and texture: sourness flavor, metallic flavor, cloying flavor, sulfurous flavor, fermented flavor, hardness, juiciness, and tenderness.

The sensory evaluation of raw samples was performed on chicken fillets stored in the original packaging. The samples were first heated to room temperature. Immediately before evaluation, an opening (4 × 4 cm) was cut into the top web of the packages, and the assessors smelled the sample trays.

For heat-treated sensory evaluation, each assessor was served one piece of chicken in a triangular shape and served from the same position of the fillet throughout the whole test. Heat treatment of chicken fillets was performed in a combi oven (Electrolux Air-o-steam, Model AOS061EANQ) at +100 °C with 100% heat for 20 min (core temperature of 72 °C). Samples were served in preheated porcelain bowls with warm metal lids and placed on a heating plate in the sensory booth. All samples were served to the panel coded with a three-digit number in duplicate following a randomized block design.

All attributes were evaluated on an unstructured 15 cm line scale with labeled end points ranging from "no intensity" (1) to "high intensity" (9). Each assessor evaluated all samples at individual speed on a computer system for direct recording of data (EyeQuestion, Software Logic8 BV, Utrecht, The Netherlands). Tap water and unsalted crackers were available for palate cleansing.

2.4.5. Statistical Analyses

The experiments were prepared using balanced experimental designs for easy analysis of the packaging choices under investigation. Subsequent data analysis was performed using analysis of variance (ANOVA) with type II sums of squares and proportions of explained variance. Tukey's pairwise comparisons were used to generate compact letter displays (CLDs), indicating which factor levels—e.g., different MAPs or number of absorbent pads—were not significantly different. All tests were performed using a level of significance of 0.05. The software used in the analyses was R version 4.0.4 [31] and the R package "mixlm" version 1.2.4 [32].

For sensory performance, ANOVA was conducted on the descriptive sensory data in order to identify the sensory attributes that discriminated among samples. A two-way mixed model was fitted for each of the sensory attributes, with the assessor and interaction effects considered to be random and the samples as a fixed effect. Least significance differences were calculated using Tukey's test ($p < 0.05$). The statistical software used for the sensory analysis was EyeOpenR (Logic8 BV).

3. Results and Discussion

3.1. Effect of Liquid Absorbent Pads and Gas Composition

The percentage drip loss was measured for chicken breast fillets packaged in modified atmospheres of 60% CO_2/40% N_2 and 75% O_2/25% CO_2 with 0, 1, and 2 liquid absorbent pads (Experiment 1) at selected sampling times during storage: 6 days, 14 days, and 20 days. The results presented in Figure 1 show that there is a clear positive relation regarding the number of liquid absorbent pads (i.e., increased absorbing surface) present in the packages. This implies that when increasing the number of the absorbent pads, the liquid lost from the product will increase as the absorbers draw excess liquid from the product. The tendency is evident for both packaging atmospheres (Figure 1). The measurements show a small increase in drip loss over the time of storage that is documented through these sampling times (6, 14, and 20 days) for all six sample variants. However, the most pronounced differences among the variants are already present at the first sampling time (day 6), revealing that the main part of the drip loss actually occurs during the initial days of storage. The relationship between the sample variants does not change after this.

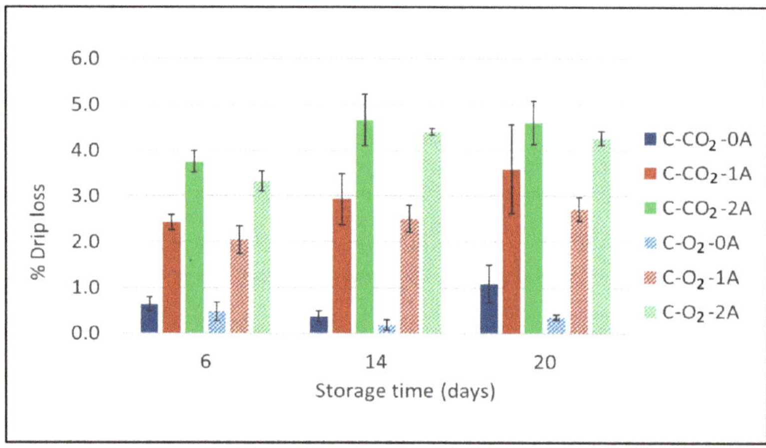

Figure 1. Drip loss for chicken breast fillets expressed as a percentage (%) of initial product weight as a function of storage time for chicken breast fillets (C) packaged in modified atmospheres of 60% CO_2/40% N_2 (C-CO_2) and 75% O_2/25% CO_2 (C-O_2) with 0, 1, or 2 liquid absorbent pads (0A, 1A, 2A). g/p ratio for all samples was 1.8 (Experiment 1). Sampling times were after 6, 14, and 20 days of cold storage. For each bar, the error bars indicate +/− one standard error.

The gas composition and storage time had significant effects on drip loss (1.8% and 3%, respectively); however, the number of absorbent pads had the most effect (88.7%) (Appendix A Table A1). Furthermore, the measured drip loss was lower for the samples packaged in 75% O_2/25% CO_2 than for those packaged in 60% CO_2/40% N_2, though not different enough to be significant. These effects were present at all sampling times (though more pronounced towards the end of storage) and for different numbers of absorbent pads. This can be explained by the solubility of CO_2 into the product, causing underpressure formation and a physical pressure on the product, resulting in increased drip loss [25,26]. The magnitude of the underpressure formed is proportional to the amount (percentage) of CO_2 present in the package, while being disproportional to the g/p ratio of the packaging concept (the available volume of the gas in relation to the volume of the product). On the other hand, increasing the CO_2 level may be beneficial in terms of improved microbiological quality—an aspect that will be considered in the following section.

In accordance with the observed increase in drip loss during storage, the dry matter content of the chicken breast fillets also increased during the initial part of storage (from day 0 to day 6), to varying extents, for all sample variants (Figure 2A). For fillets stored in a CO_2-rich atmosphere, the dry matter content of the chicken meat increased until 14 days of storage, followed by a decrease in dry matter content. For samples stored in an O_2-rich atmosphere the effect of storage time on the dry matter content was more ambiguous, and no clear correlation could be deducted from the results.

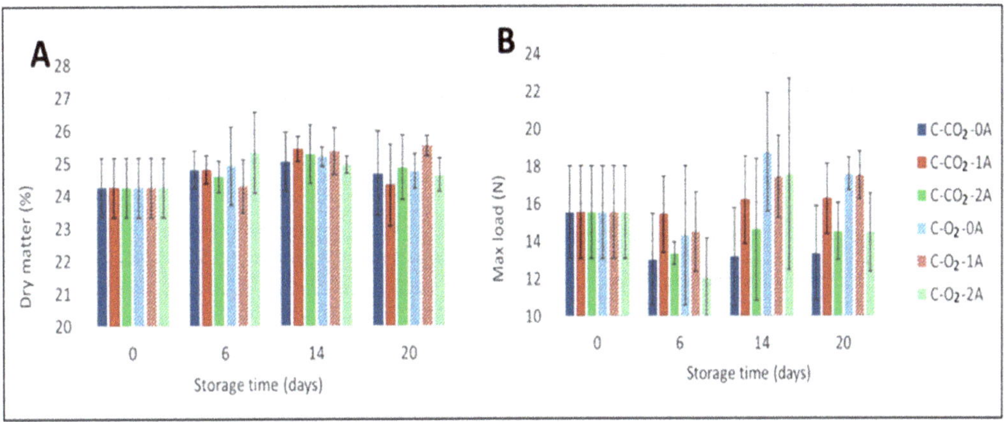

Figure 2. Percentage dry matter content measured for chicken breast fillets (**A**) (two replicates per sample variant) and Warner–Bratzler maximum shear force (N/cm^2) (WBSF) required to cut cooked samples of chicken breast fillets (**B**) (presented as an average of 6 replicates per sample variant). The chicken breast fillets (C) were stored in modified atmospheres of 60% CO_2/40% N_2 (C-CO_2) and 75% O_2/25% CO_2 (C-O_2) (g/p 1.8) with 0, 1, or 2 liquid absorbent pads (0A, 1A, 2A) (Experiment 1) and sampling times of 0, 6, 14, and 20 days.

In general, for all samples there was a tendency for the dry matter content to increase from the time of packaging to the end of storage, but the dry matter content varied among the sample variants and by the time of storage. Regarding the number of liquid absorbent pads, no clear correlation between an increased number of pads and dry matter content could be documented—neither for the samples in CO_2-rich atmospheres, nor for those in O_2-rich atmospheres. Correlation analyses showed that the correlation between dry matter and drip loss was −0.166 in CO_2-rich atmospheres, while no correlation (−0.001) was observed in O_2-rich atmospheres. Furthermore, the number of pads had no significant effect as a main factor (Appendix A Table A1).

The effects of the number of liquid absorbent pads and of gas composition were also evaluated through texture analysis using the Warner–Bratzler method. As displayed in Figure 2B, for fillets stored in CO_2-rich atmospheres the measured WBSF shows a net decrease from the time of packaging (15.6 N/cm^2) to the end of storage for samples stored without absorbent pads (0A) (13.4 N/cm^2) and for those stored with two absorbent pads (2A) (14.5 N/cm^2). For fillets stored with one absorbent pad (1A) the measured WBSF was practically unchanged throughout storage. In general, the measured differences between 0, 1, and 2 absorbent pads for this gas composition were very small. In addition, according to one-way ANOVA—computed by defining a six-level factor with the combinations of gas composition and number of liquid absorbent pads—there were no significant differences between these samples at any storage time (Appendix A Table A2). Although all main factors (gas composition, number of absorbent pads, and storage time) were significant (Appendix A Table A1), the variance explained by these factors was relatively low (7.2%, 6.6%, and 13.2%, respectively), leaving 59% as unexplained and a corresponding R^2 of 41%, i.e., other sources of variation dominated the WBSF.

For storage in high O_2-atmospheres, an initial reduction in maximum force was measured for all three numbers of absorbent pads (0, 1, or 2 absorbent pads) added. Perhaps most interesting in these results is the increase in measured maximum force between day 6 and day 14 of storage, indicating a decrease in tenderness of the fillets cooked after this time of storage, seen only for high-O_2 samples. The measured shear force dropped at the last sampling day (day 20) of storage regardless of the number of absorbent pads. For fillets stored zero (0A) or with one absorbent pad (1A), the texture evaluated by WBSF was measured to be higher (17.2 N/cm^2 and 17.5 N/cm^2) at the end of storage compared to the initial level (15.6 N/cm^2), indicating a decrease in meat tenderness towards the end of storage. For chicken stored with two absorbent pads (2A), the measured WBSF was reduced at the end of storage (14.5 N/cm^2) compared to the start (15.6 N/cm^2). However, according to one-way ANOVA, no significant differences between the samples stored in high-O_2 atmospheres were detected at any sampling time. The only significant differences were observed between samples stored in different atmospheres after 20 days of storage (Appendix A Table A2). The physicochemical origin of the observed differences between samples stored in high-CO_2 and high-O_2 atmospheres is unknown. For differences in tenderness between chicken fillets to be of importance for the perceived quality, the numeric differences in measured shear force would assumedly need to be a great deal larger. Again, even though the fillet selection was randomized at packaging, some of the measured variations may be a result of individual variations.

The results from sensory evaluation of the textural traits hardness, tenderness, and juiciness of heat-treated chicken breast fillets (Figure 3) reveal that there are significant differences in the perceived texture of the meat as affected by the number of absorbent pads (flavor evaluation of heat-treated chicken breast fillets and odor evaluation of raw chicken are presented in Appendix A Table A3, and will be discussed in relation to microbial growth).

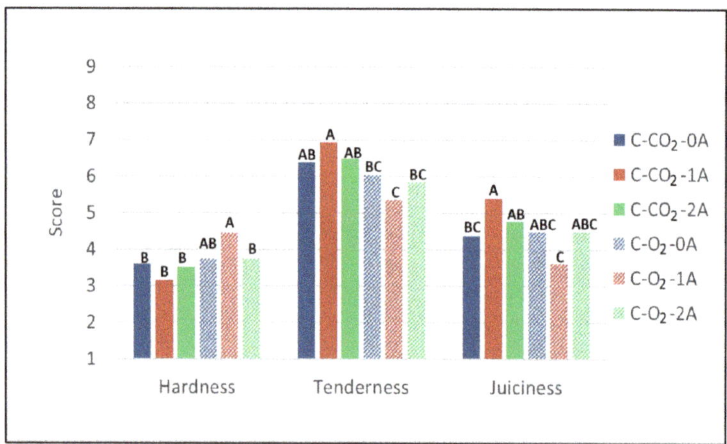

Figure 3. Sensory intensity scores (scale 1–9) for the textural attributes hardness, tenderness, and juiciness of heat-treated chicken breast fillets (C) stored for 14 days in modified atmospheres of 60% CO_2/40% N_2 (C-CO_2) and 75% O_2/25% CO_2 (C-O_2) with 0, 1, or 2 liquid absorbent pads (0A, 1A, 2A). g/p ratio for all samples was 1.8 (Experiment 1).

However, the trend is not systematic across gas compositions, as it does not increase/decrease with the increasing number of pads independent of the gas composition, i.e., the only significant difference is found in the interaction between the number of absorbent pads and the packaging gas. A significantly higher score of hardness with one absorbent pad (1A) compared to two absorbent pads (2A) was observed for samples stored in high-O_2 atmospheres, while regarding the juiciness of chicken stored in high-CO_2 atmospheres, a lower score for samples stored without absorbent pads (0A) compared to one absorbent pad (1A) was observed. In this experiment, tenderness and juiciness both correlated positively with dry matter content (0.231 and 0.210, respectively) and negatively with WBSF (−0.517 and −0.197, respectively), while hardness did not correlate with dry matter (−0.040) but correlated positively with WBSF (0.397). However, although there were significant differences in sensory scores among different numbers of absorbent pads, the differences were less than 2 in score (hardness 3.75 and 4.45; juiciness 4.37 and 5.40), i.e., less than what is presumed observable by consumers. In addition, no significant differences in dry content or texture measured as maximum force (WBSF) (neither number of absorbent pads nor gas composition) were observed for the samples stored for 14 days. This implies that for this product consumers will not be able to identify differences in the textural quality of fillets packaged with liquid absorbent pads overdimensioned for the product's drip loss.

With regard to the effect of modified atmosphere composition on the textural attributes, the samples stored in high-O_2 atmospheres were evaluated with slightly higher sensory intensity scores for hardness than the ones stored in high-CO_2 atmospheres. This is supported by the fact that the sample variants stored in high-O_2 atmospheres were evaluated to have somewhat lower intensity scores on tenderness than the sample variants stored in high-CO_2 atmospheres. For juiciness—an important quality trait for chicken breast fillets—no clear tendency can be observed when comparing the sample variants stored in high-O_2 compared to high-CO_2 modified atmospheres. However, the gas composition significantly affected the texture in samples with one absorbent pad (1A), as shown by lower scores of tenderness and juiciness and higher scores of hardness in the samples stored in high-O_2 atmospheres compared to high-CO_2 atmospheres. This finding is in accordance with the results of the Warner–Bratzler shear force measurements, which, as discussed, displayed the largest relative differences between the two gas compositions at measurement after 14 days of storage—the same storage time at which sensory evaluation was performed. Still, the differences in sensory scores are small and will most likely not be detectable by the average consumer. Geesink et al. (2015) stated that the effect of high oxygen on tenderness of meat has been reported in a number of studies [33]. To the best of our knowledge, reported effects of packaging atmosphere on textural attributes of chicken/poultry meat are rather scarce. Rossaint et al. (2015) reported no significant difference in texture (measured as sensory attribute) for poultry stored in a high-oxygen atmosphere (70% O_2/30% CO_2) compared to 70% N_2/30% CO_2 nitrogen [34].

Regarding texture measured as maximum force (WBSF), a majority of publications have been on beef and pork meat, and often conducted comparisons of high-oxygen atmospheres with vacuum (oxygen- and CO_2-free atmosphere) [35–38]. Lagersted et al. (2011) reported that storage in MAP with high oxygen resulted in higher shear force and negatively affected the juiciness and tenderness of beef steaks compared to vacuum [37]. Moczkowska et al. (2017) reported a decrease in shear force (WBSF) for beef stored in an oxygen-free atmosphere (vacuum), while increased WBSF was observed when stored in a high-oxygen atmosphere [36]. Similarly, Zakrys-Waliwander et al. (2012) detected significantly lower WBSF in beef steaks stored in vacuum compared to a high-O_2 atmosphere after 8 and 14 days, but not after 4 days of storage [38]. For pork meat (porcine longissimus dorsi) a decrease in tenderness was detected already after 4 days of storage in a high-oxygen atmosphere, with further decrease until 14 days of storage, compared to an increase in tenderness when stored in an oxygen-free atmosphere (vacuum) [39]. In oxygen-rich atmospheres the potential for increased oxidation is present, and oxidation of protein can influence properties such as water-holding capacity, and lead to changes in texture, such as tenderness [36,39].

The TVC at the start of storage (day 0) was approximately 2.6 log cfu/g for all sample variants (Appendix A Table A4). In general, the increase in TVC developed in a parallel manner for sample variants in high-O_2 and high-CO_2 atmospheres, with a slightly lower bacterial growth on the samples in high-CO_2 atmospheres, reaching 6.7–7.1 log cfu/g after 20 days of storage for the high-CO_2 variants, and 7.3–7.5 log cfu/g for the high-O_2 atmosphere samples. Overall, there was a small but significant effect of gas composition on TVC (Appendix A Table A4). Previous results show similar total viable count numbers for products packaged with and without high levels of oxygen (though with low CO_2-levels; 30% CO_2/70% N_2 and 30% CO_2/70% O_2) [40], whereas in the present study high levels of CO_2 resulted in lower TVCs compared to high-O_2 atmospheres. Statistical analysis also confirmed that gas composition had a significant effect on the TVC (although only 1.7% explained variance), while there was no significant effect of different numbers of absorbent pads on TVC. Hence, the microbiological growth is independent of whether or not there is free liquid in the packages.

This finding was further supported by results from additional analysis performed at day 14 of storage by sampling from the bottom surfaces of the fillets that were in contact with available visible liquid and comparing the TVCs measured in these samples to the TVCs from the top surfaces of the fillets at the same sampling day (Appendix A Table A4). The TVC data for the top and bottom samples for all six packaging variants are summarized in Appendix A Table A4. For samples stored in high-CO_2 atmospheres, a significantly lower bacterial level was measured for samples taken from the bottom surface (and in contact with visible liquid if present) compared to samples from the top surface of the fillets (with two absorbent pads (2A)). However, no such differences in bacterial growth for samples stored in high-O_2 atmospheres with different numbers of absorbent pads could be detected. Hence, free liquid inside packages does not give rise to increased microbiological activity. Even in the packaging variants without an absorbent pad present (0A), the microbiological growth at the bottom and top of the fillets was similar—not significantly different—for the same packaging conditions, as well as similar to the measured TVC levels for packaging with the highest number of absorbent pads (no free liquid inside the packages). This is a significant finding, as it disproves the common perception that packaged meat with excess liquid inside the packages has a poorer microbial quality [18]. Dissemination of these results to the food industry, retailers, and consumers could contribute to reducing food waste in this product category.

Regardless of the number of absorbent pads added, growth of *Brochothrix thermosphacta* reaches a level of approximately 7 log cfu/g for fillets in high-O_2 atmospheres and 3–4 log cfu/g in high-CO_2 atmospheres after 20 days of storage (Appendix A Table A4). This is as expected for this bacterium; *Brochothrix* grows fast in the presence of O_2, and it is also able to adapt to an anaerobic environment, though then at a much slower growth rate, which is in accordance with previous studies [34,41].

Lactic acid bacteria increased from 2.45 log cfu/g to approximately 7 log cfu/g (7.12–7.31 log cfu/g and 6.67–7.22 log cfu/g in high-CO_2 and high-O_2 atmospheres, respectively) over a 20-day storage period. Levels of *Enterobacteriaceae* were measured to be about 2 log cfu/g until 15 days of storage, followed by an increase to approximately 3 log cfu/g in high-O_2 atmospheres (2.23–3.17 log cfu/g), and slightly higher—approximately 4 log cfu/g—in high-CO_2 atmospheres (4.02–4.27 log cfu/g) after 20 days of storage. According to analyses of variance, the number of liquid absorbent pads had no significant effect on the levels of lactic acid bacteria and *Enterobacteriaceae*. No significant differences were observed at any sampling time within each packaging gas (Appendix A Table A4), and the variance was mainly explained by the storage time (Appendix A Table A1). Moreover, no significant effect of the number of absorbent pads in the packages on the growth of *B. thermosphacta* could be seen in these results, but in this case most of the variance was explained by gas composition and storage time (45.4% and 42%, respectively).

Sensory evaluation of the odor of the raw chicken showed similar trends as the bacterial growth, with no significant effect of different numbers of liquid absorbent pads but clear effect of the gas composition (Appendix A Table A3). In all evaluated attributes, significant differences were detected, with slightly higher scores for sourness and metallic odors and slightly lower scores for chlorine odor in chicken stored in high-CO_2 atmospheres compared to high-O_2 atmospheres. However, for the other attributes, the differences were more pronounced. Chicken stored in high-O_2 atmospheres had high scores (6.88–7.28) for fermented and cloying (6.48–6.89) odors compared to storage in high-CO_2 atmospheres (2.14–2.87 and 3.37–3.94, respectively). The opposite was the case regarding the sulfurous odor: 1.97–2.07 in high-O_2 atmospheres and 5.38–6.23 in high-CO_2 atmospheres. For heat-treated chicken the only significant flavor attributes were sourness and fermented flavors, though the scores were relatively low (below 3.37) and the difference was less than 2 units.

3.2. Effect of Liquid Absorbent Pads and G/P Ratio

The g/p ratio applied in the first experiment in the study (g/p 1.8) is comparable to what is commonly used for modified atmosphere packaging on the market in Norway. Based on the findings of this first and main part of the study, it was relevant to look into the effect of varying the g/p ratio, i.e., the amount of carbon dioxide gas available to the product, and how it affects the drip loss and microbiological quality of the chicken breast fillets when packaged with 0, 1, or 2 liquid absorbent pads. Therefore, the second part of the study includes both g/p 1.8 (as for Experiment 1) and a higher g/p ratio of 2.9.

Carbon dioxide has a well-documented antimicrobial effect [13,42–44]. According to Renerre et al. (1990), Dalgaard et al. (1993), and Randell et al. (1999), the drip loss increases with increased CO_2 content in the packages [20–22]. However, Holck et al. (2014) showed that drip loss is not solely dependent on the amount of CO_2, with higher drip loss in chicken packed in 100% CO_2 compared to 100% CO_2 with the addition of a CO_2 emitter [25]. Thus, underpressure formation with high amounts of CO_2 in rigid packages may result in an excessive drip loss from the product, due to physical pressure as the package is compressed [26]. Figure 4 displays the drip loss results from the second experiment on chicken breast fillets. Firstly, this follow-up experiment confirms the results obtained in the initial experiment: the drip loss increases when increasing the number of liquid absorbent pads. After 20 days of storage, two absorbent pads (2A) resulted in significantly higher drip loss compared to without absorbent pads (0A). However, after 7 and 14 days of storage, significantly higher drip loss was detected in packages with one absorbent pad (1A) compared to those without absorbent pads (0A) at g/p 1.8 (Appendix A Table A2). Secondly, the figure shows that there is a general tendency towards higher drip loss for samples in modified atmospheres at g/p 1.8 compared to g/p 2.9. By statistical analysis, this effect was found to be significant, with higher drip loss at g/p 1.8 compared to g/p 2.9 with one absorbent pad (1A) after both 7 and 14 days of storage. Analysis of variance shows that all main factors (g/p ratio, number of absorbent pads and storage time) were significant (Appendix A Table A1), with explained variances of 16.7%, 26.1%, and 31.6%, respectively. At a higher g/p ratio (and identical gas composition) there is a larger amount of CO_2 present in the package and, thus, a higher amount of CO_2 that can be dissolved into the product. This implies that when CO_2 is absorbed by the product the underpressure formed will be relatively small in the packages with g/p 2.9 compared to in the packages with g/p 1.8. The samples with g/p 2.9 consequently have a lower drip loss due to a less pronounced physical pressure on the chicken breast fillets. The variance in the measurements reflects the natural individual variations between the products.

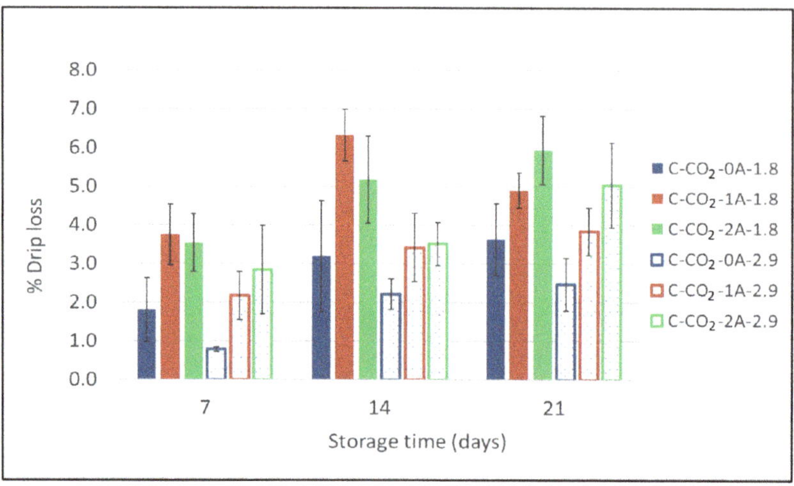

Figure 4. Drip loss for chicken breast fillets expressed as a percentage (%) of initial product weight as a function of storage time for chicken breast fillets (C) packaged in modified atmospheres of 60% CO_2/40% N_2 (C-CO_2) with 0, 1, or 2 liquid absorbent pads (0A, 1A, 2A) and two different g/p ratios (1.8 and 2.9) (Experiment 2). Sampling times were after 7, 14, and 20 days of cold storage.

Statistical analysis of variance showed that the main differences in TVC were explained by storage time (98.4%) (Appendix A Table A1). Some differences in TVC were detected after 14 days of storage, but there were no significant differences for samples stored for 20 days, showing similar bacterial counts as for the previous TVC results (Appendix A Table A4). This indicates that the microbial growth on the chicken breast fillets is not affected by changing the g/p ratio—and thereby, the amount of available CO_2—for this packaging concept. Due to the antimicrobial effects of CO_2, increasing the g/p ratio may have been expected to produce an increased antimicrobial effect of the packaging gas. However, the percentage of CO_2 in the headspace was the same, and the results demonstrate that the amount of CO_2 available at a g/p of 1.8 is adequate to produce a comparable effect of CO_2 in terms of microbiological quality as measured for a g/p of 2.9. These results also support the finding that microbiological growth on the chicken fillets is independent of the presence of or the number of liquid absorbent pads in the packages, i.e., whether or not there is free liquid inside the packages.

4. Conclusions

A significant increase in drip loss with an increasing number of liquid absorbent pads was documented for chicken breast fillets. The percentage of dry matter content increased during storage; however, only minor effects of gas composition and the number of absorbent pads could be detected. Sensory properties related to the texture were not significantly affected by the number of pads for the fillets. Microbiological analyses showed that TVC was independent of the number of absorbent pads, i.e., whether or not there was free liquid in the packages. This is an interesting finding as it disproves the common belief that excess liquid inside packages results in a higher bacterial load on the product. By comparing two different packaging gases in MAP, it was confirmed that for both packaging gases increased numbers of liquid absorbent pads in the packages led to increased drip loss. Still, there was no pronounced effect on sensory quality (texture), though regardless of liquid absorption, higher intensity scores of negatively associated odor attributes were detected for the high-O_2 atmospheres compared to the high-CO_2 atmospheres.

This study demonstrates the importance of product-adjusted capacity of liquid absorbers in order to maintain product yield, which can have economic benefits for food producers and retailers. Visible free liquid inside the packages did not affect the bacterial load, but might result in rejection by the consumer due to unattractive packages, which might have an effect in terms of increased food waste.

Author Contributions: Conceptualization, M.K.P.; methodology, M.K.P. and M.C.; validation, M.K.P., J.N.-N., A.Å.H., M.C. and K.H.L.; formal analysis, K.H.L. and M.K.P.; investigation, M.K.P., A.Å.H., and J.N.-N.; writing—original draft preparation, M.K.P. and J.N.-N.; writing—review and editing, M.K.P., J.N.-N., A.Å.H., M.C. and K.H.L.; visualization, M.K.P. and J.N.-N. All authors have read and agreed to the published version of the manuscript.

Funding: This work was funded by the Norwegian Levy on Agricultural Products through the Strategic Programs FoodMicro-Pack (Project no 262306) and FutureFoodControl (Project no 314743), and The Research Council of Norway in the "FoodPilot Plant (Project no 296083).

Informed Consent Statement: Informed consent was obtained from all subjects involved in the study.

Data Availability Statement: The data presented in this study are available on request from the corresponding author.

Acknowledgments: We would like to thank Aud Espedal, Janina Berg, Karen Wahlstrøm Sanden, and Jens Petter Wold for their skillful technical assistance.

Conflicts of Interest: The authors declare no conflict of interest.

Appendix A

Table A1. Summaries of analyses of variance for measurements on chicken breast fillets (Experiments 1 and 2). Numbers show percentage of variation per factor/interaction.

Chicken Breast Fillet (Experiment 1): G/P 1.8							
Factor/Effects	Drip Loss	Total Viable Counts	Lactic Acid Bacteria	*Brochothrix thermosphacta*	*Enterobacteriaceae*	Max Load	Dry Matter Content
Gas composition (G)	1.8 ***	1.7 ***	0	45.4 ***	5.2 ***	7.2 **	0.6
Storage time (T)	3.0 ***	95.7 ***	95.4 ***	42.0 ***	58.0 ***	13.2 **	6.4
Liquid absorbent pads (A)	88.7 ***	0	0.3 °	0.3	0.6	6.6 *	0.1
G × T	0.3	-	0.9 ***	4.1 ***	9.5 ***	6.3 °	-
G × A	0.1	-	0.3 *	0.8 *	0.9	6.5 *	-
T × A	2.0 ***	-	0.4 °	1.5 **	2.2	1.2	-
Residuals (Error)	4.1	2.6	2.7	5.9	23.6	59.0	92.9
R^2	95.8	97.4	97.3	94.1	76.4	41.0	7.1
Chicken Breast Fillet (Experiment 2): Gas Composition: 60% CO_2/40% N_2							
	Drip Loss	Total Viable Counts	Lactic Acid Bacteria	*Brochothrix thermosphacta*	*Enterobacteriaceae*		
Gas volume/product volume ratio (GP)	16.7 ***	0.1 °	0.1 *	5.7 ***	8.1 ***		
Storage time (T)	26.1 ***	98.4 ***	97.3 ***	71.7 ***	66.5 ***		
Liquid absorbent pads (A)	31.6 ***	0	0.2 *	1.1 °	2.5 *		
GP × T	-	0	0	6.7 ***	0		
GP × A	-	0.2 *	0.5 ***	0.4	3.5 **		
T × A	-	0.2 °	0.9 ***	2.8 *	0.3		
Residuals (Error)	25.6	1.1	1.0	11.7	19.1		
R^2	74.1	98.9	99.0	88.4	80.9		

p-values are indicated as 0 *** 0.001 ** 0.01 * 0.05 ° 0.1 e.g., one star for *p*-values between 0.01 and 0.05.

Table A2. One-way analyses of variance for drip loss, dry matter content and Warner–Bratzler shear force (N/cm^2) (maximum load) for each recorded day of storage of chicken breast fillets (Experiment 1). The one-way ANOVAs were computed by defining a six-level factor with the combinations of gas composition and liquid absorbent pads. Tukey's letters signify which of the six combinations of gas composition and liquid absorbent pad numbers are not significantly different (p-value > 0.05), i.e., levels sharing a letter are not significantly different. In the lower part of the table (Experiment 2), gas volume/product volume replaces gas composition.

Chicken Breast Fillet (Experiment 1): G/P 1.8						
	Gas Composition	**Number of Liquid Absorbent Pads**	**0 Days**	**6 Days**	**14 Days**	**20 Days**
Drip loss	60% CO_2/40% N_2	0	-	0.65 C	0.37 C	1.09 C
		1	-	2.42 B	2.93 B	3.60 AB
		2	-	3.76 A	4.67 A	4.61 A
	25% CO_2/75% O_2	0	-	0.48 C	0.18 C	0.35 C
		1	-	2.05 B	2.51 B	2.72 B
		2	-	3.33 A	4.41 A	4.27 A
Dry matter Content	60% CO_2/40% N_2	c	24.25	24.81 A	25.05 A	24.68 A
		1	24.25	24.80 A	25.44 A	24.32 A
		2	24.25	24.59 A	25.28 A	24.87 A
	25% CO_2/75% O_2	0	24.25	24.91 A	25.20 A	24.75 A
		1	24.25	24.29 A	25.37 A	25.52 A
		2	24.25	25.32 A	24.94 A	24.63 A
Warner–Bratzler shear force (maximum load)	60% CO_2/40% N_2	0	15.56	13.02 A	13.17 A	13.35 B
		1	15.56	15.46 A	16.21 A	16.27 AB
		2	15.56	13.35 A	14.63 A	14.53 AB
	25% CO_2/75% O_2	0	15.56	14.3 A	18.74 A	17.59 A
		1	15.56	14.51 A	17.44 A	17.52 A
		2	15.56	12.02 A	17.58 A	14.47 AB
Chicken Breast Fillet (Experiment 2): Gas Composition: 60% CO_2/40% N_2						
	Gas Volume/ Product Volume	**Number of Liquid Absorbent Pads**	**0 Days**	**7 Days**	**14 Days**	**20 Days**
Drip loss	1.8	0	-	1.81 BC	3.20 BC	3.63 BC
		1	-	3.64 A	6.05 A	5.12 AB
		2	-	3.54 AB	5.17 AB	5.95 A
	2.9	0	-	0.78 C	2.22 C	2.47 C
		1	-	2.18 ABC	3.42 BC	3.84 BC
		2	-	2.84 AB	3.51 BC	5.04 AB

Table A3. Sensory intensity scores (scale 1–9) for flavor attributes for heat-treated samples and odor attributes for raw chicken breast fillets stored for 14 days in modified atmospheres of 60% CO_2/40% N_2 and 75% O_2/25% CO_2 with 0, 1, or 2 liquid absorbent pads (Experiment 1). Levels sharing a letter are not significantly different (p < 0.05) between the chicken breasts samples, as measured by two-way ANOVA and Tukey's multiple comparison test.

Flavor of Heat-Treated Chicken—14 Days of Storage									
Packaging Variable: Gas Composition	**Number of Liquid Absorbent Pads**	**Sourness**	**Metallic**	**Cloying**	**Sulfurous**	**Fermented**	**Hardness**	**Tenderness**	**Juiciness**
60%CO_2/40% N_2	0	2.87 AB	4.23	3.88	2.89	1.71 B	3.59 B	6.38 AB	4.37 BC
	1	2.89 AB	4.19	4.08	2.61	1.89 B	3.16 B	6.92 A	5.40 A
	2	3.26 A	4.38	3.38	2.35	1.48 B	3.50 B	6.49 AB	4.77 AB
25%CO_2/75% O_2	0	3.13 A	3.79	3.67	2.73	2.43 AB	3.75 AB	6.04 BC	4.49 ABC
	1	1.80 B	3.56	4.55	3.25	3.37 A	4.45 A	5.35 C	4.63 C
	2	2.39 AB	3.63	4.39	2.96	2.59 AB	3.74 B	5.84 BC	4.48 ABC
p-value		0.0058	0.0556	0.1608	0.2917	0.0008	0.0001	0.0000	0.0001

Table A3. *Cont.*

		Odor of Raw Chicken—14 Days of Storage					
		Sourness	Metallic	Cloying	Sulfurous	Fermented	Chlorine
60% CO_2/40% N_2	0	2.58 A	4.42 A	3.66 B	6.23 A	2.36 B	1.35 B
	1	2.87 A	4.39 A	3.37 B	5.38 A	2.14 B	1.31 B
	2	2.40 A	4.46 A	3.94 B	5.62 A	2.87 B	1.28 B
25% CO_2/75% O_2	0	1.23 B	2.26 B	6.89 A	2.07 B	7.24 A	3.41 A
	1	1.30 B	2.28 B	6.81 A	2.07 B	7.28 A	3.48 A
	2	1.24 B	2.42 B	6.48 A	1.97 B	6.88 A	3.62 A
p-value		0.0000	0.0000	0.0000	0.0000	0.0000	0.0000

Table A4. Microbiological content measured as total viable count of bacteria (TVC), lactic acid bacteria (LAB), *Brochothrix thermosphacta,* and *Enterobacteriaceae* given as log cfu/g for chicken breast fillets (Experiment 1 and 2) as a function of storage time, packaging variables, gas composition (60% CO_2/40% N_2 and 25% CO_2/75% O_2), or g/p ratio (1.8, 2.9, 2.0 and 2.1), with 0, 1, or 2 liquid absorbent pads. Sampling times after 6 (7), 14, and 20 days of cold storage. Values are mean ± standard deviation (n = 3).

Storage Time (Days)	Packaging Variable	Number of Liquid Absorbent Pads	Total Viable Counts	Lactic Acid Bacteria	*Brochothrix thermosphacta*	*Enterobacteriaceae*
		Chicken Breast Fillet (Experiment 1): G/P 1.8				
	Gas Composition					
0			2.64 ± 0.26	2.45 ± 0.36	2.0 ± 0.0 **	2.64 ± 0.13 **
6	60% CO_2/40% N_2	0	3.04 ± 0.28 B	2.83 ± 0.39 B	1.50 ± 0.39 ** C	1.63 ± 0.41 ** A
		1	3.54 ± 0.19 AB	2.73 ± 0.25 A	1.42 ± 0.24 ** C	1.42 ± 0.24 ** A
		2	3.52 ± 0.29 AB	3.61 ± 0.32 A	1.42 ± 0.24 C	1.82 ± 0.46 A
	25% CO_2/75% O_2	0	3.83 ± 0.14 A	3.87 ± 0.16 A	3.48 ± 0.35 A	1.99 ± 0.11 A
		1	3.72 ± 0.17 A	3.65 ± 0.21 A	3.29 ± 0.40 AB	1.82 ± 0.45 A
		2	3.56 ± 0.30 AB	3.59 ± 0.41 A	2.22 ± 1.07 ** BC	1.68 ± 0.50 A
14	60% CO_2/40% N_2	0	5.59 ± 0.09 AB	5.78 ± 0.0 * A	2.50 ± 0.81 B	2.10 ± 0.98 BC
		0 (Bo)	5.65 ± 0.38 AB	5.70 ± 0.0 * AB	2.44 ± 0.45B	3.27 ± 0.16 AB
		1	5.71 ± 0.0 AB	5.70 ± 0.0 * AB	2.10 ± 0.0 B	1.96 ± 0.0 BC
		1 (Bo)	5.50 ± 0.41 B	5.60 ± 0.0 ** AB	1.68 ± 0.45B	2.07 ± 0.75 BC
		2	5.55 ± 0.30 AB	5.60 ± 0.0 * AB	2.93 ± 0.59 B	1.72 ± 0.68 BC
		2 (Bo)	4.69 ± 0.40 C	4.85 ± 0.32 C	2.04 ± 1.11B	1.0 ± 0.0 ** C
	25% CO_2/75% O_2	0	6.06 ± 0.15 AB	5.60 ± 0.0 * AB	5.45 ± 0.30 * A	1.83 ± 0.81 BC
		0 (Bo)	6.23 ± 0.08 A	5.78 ± 0.0 * A	5.24 ± 0.19 A	3.86 ± 0.32 A
		1	5.84 ± 0.30 AB	5.46 ± 0.17 B	5.70 ± 0.0 * A	1.94 ± 0.66 BC
		1 (Bo)	5.86 ± 0.20 AB	5.78 ± 0.0 * A	5.40 ± 0.15A	2.57 ± 1.36ABC
		2	5.96 ± 0.32 AB	5.63 ± 0.05 * AB	5.61 ± 0.07 * A	1.12 ± 0.24 C
		2 (Bo)	5.56 ± 0.29 AB	5.51 ± 0.06 * B	5.13 ± 0.09A	1.12 ± 0.34 ** C
20	60% CO_2/40% N_2	0	6.94 ± 0.09 BC	7.12 ± 0.14 AB	4.08 ± 0.62 B	4.27 ± 0.87 * A
		1	6.66 ± 0.33 C	7.31 ± 0.43 A	2.92 ± 0.59 C	4.13 ± 0.12 A
		2	7.07 ± 0.18 ABC	7.31 ± 0.25 A	3.99 ± 0.72 BC	4.02 ± 0.07 A
	25% CO_2/75% O_2	0	7.29 ± 0.22 AB	6.66 ± 0.08 B	6.86 ± 0.39 A	2.28 ± 0.96 B
		1	7.36 ± 0.21 AB	6.80 ± 0.16 AB	6.93 ± 0.15 A	3.17 ± 0.46 AB
		2	7.53 ± 0.25 A	7.22 ± 0.25 A	7.40 ± 0.36 A	2.91 ± 0.61 AB

Table A4. *Cont.*

Chicken Breast Fillet (Experiment 2): Gas Composition: 60% CO_2/40% N_2						
Storage Time (Days)	Packaging Variable	Number of Liquid Absorbent Pads	Total Viable Counts	Lactic Acid Bacteria	*Brochothrix thermosphacta*	*Enterobacteriaceae*
	Gas/Product Volume Ratio					
0			3.30 ± 0.41	2.11 ± 0.31	1.55 ± 0.24	1.30 ± 0.0
7	1.8	0	3.48 ± 0.04 ABC	3.65 ± 0.05 AB	1.30 ± 0.0 ** A	1.77 ± 0.38 A
		1	3.41 ± 0.12 BC	3.41 ± 0.17 AB	1.30 ± 0.0 ** A	1.89 ± 0.68 ** A
		2	3.80 ± 0.29 A	3.74 ± 0.28 A	1.38 ± 0.15 A	2.10 ± 0.43 A
	2.9	0	3.53 ± 0.09 ABC	3.49 ± 0.08 AB	1.45 ± 0.30 ** A	1.30 ± 0.0 ** A
		1	3.82 ± 0.16 AB	3.68 ± 0.12 AB	1.30 ± 0.0 ** A	1.50 ± 0.24 ** A
		2	3.29 ± 0.12 C	3.32 ± 0.13 B	1.30 ± 0.0 ** A	1.30 ± 0.0 ** A
14	1.8	0	6.22 ± 0.1 B	6.30 ± 0.0 *	2.18 ± 0.67 C	3.95 ± 0.13 A
		1	6.42 ± 0.10 A	6.30 ± 0.0 *	2.72 ± 0.57 BC	3.94 ± 0.33 A
		2	6.43 ± 0.04 A	6.30± 0.0 *	2.30 ± 0.61 C	3.75 ± 0.45 AB
	2.9	0	6.43 ± 0.02 A	6.54± 0.0 *	3.20 ± 1.12ABC	2.45 ± 1.10 B
		1	6.40 ± 0.01 A	6.54± 0.0 *	4.57 ± 0.66 A	3.37 ± 0.59 AB
		2	6.00 ± 0.0 * C	5.48± 0.0	4.02 ± 0.04 AB	3.72 ± 0.35 AB
20	1.8	0	7.34 ± 0.16 A	7.29 ± 0.11 AB	3.68 ± 0.29 A	4.22 ± 0.34 A
		1	7.35 ± 0.15 A	7.20 ± 0.10 AB	3.09 ± 0.79 A	4.03 ± 0.63 A
		2	7.40 ± 0.19 A	7.38 ± 0.11 A	4.34 ± 0.35 A	3.71 ± 0.47 AB
	2.9	0	7.13 ± 0.16 A	7.00 ± 0.11 B	3.94 ± 0.48 A	2.69 ± 0.81 B
		1	7.20 ± 0.16 A	7.27 ± 0.09 AB	4.00 ± 0.0 ** A	3.48 ± 0.68 AB
		2	7.53 ± 0.33 A	7.41 ± 0.23 A	4.36 ± 0.85 A	4.11 ± 0.61 A

(Bo) = sampling from the bottom surfaces of the chicken breast fillets. (*) value given as > (too low dilutions applied—at least two replicates over growth and not possible to count); (**) value given as < (too high dilutions applied–at least two replicates below detection limit at this dilution). Means sharing letters (Tukey) within the same column and storage time are not significantly different ($p > 0.05$) (one-way ANOVA with six levels per day).

References

1. United Nations. Sustainable Development Goals, Transforming Our World: The 2030 Agenda for Sustainable Development. Available online: https://sustainabledevelopment.un.org/post2015/transformingourworld (accessed on 15 January 2021).
2. Ellen MacArthur Foundation. The New Plastics Economy: Rethinking the Future of Plastics and Catalysing Action. Available online: https://www.ellenmacarthurfoundation.org/assets/downloads/publications/NPEC-Hybrid_English_22-11-17_Digital.pdf (accessed on 2 May 2021).
3. Caldeira, C.P.; De Laurentiis, V.; Sala, S. *Assessment of Food Waste Prevention Actions. Development of an Evaluation Framework to Assess the Performance of Food Waste Prevention Actions*; EUR 29901 EN; JRC: Luxemburg, 2019.
4. Lindh, H.; Williams, H.; Olsson, A.; Wikström, F. Elucidating the Indirect Contributions of Packaging to Sustainable Development: A Terminology of Packaging Functions and Features. *Packag. Technol. Sci.* **2016**, *29*, 225–246. [CrossRef]
5. Svanes, E.; Vold, M.; Møller, H.; Pettersen, M.K.; Larsen, H.; Hanssen, O.J. Sustainable packaging design: A holistic methodology for packaging design. *Packag. Technol. Sci.* **2010**, *23*, 161–175. [CrossRef]
6. Verghese, K.; Crossin, E.; Clune, S.; Lockrey, S.; Williams, H.; Rio, M.; Wikström, F. The greenhouse gas profile of a "Hungry Planet"; quantifying the impacts of the weekly food purchases including associated packaging and food waste of three families. In Proceedings of the 19th IAPRI World Conference, Melbourne, Australia, 15–18 June 2014.
7. Wikström, F.; Verghese, K.; Auras, R.; Olsson, A.; Williams, H.; Wever, R.; Grönman, K.; Kvalvåg Pettersen, M.; Møller, H.; Soukka, R. Packaging Strategies That Save Food: A Research Agenda for 2030. *J. Ind. Ecol.* **2019**, *23*, 532–540. [CrossRef]
8. Prestrud, K.; Stensgård, A.E.; Møller, H.; Johnsen, F.M. *Emballasjeutviklingen i Norge 2017. Handlekurv og Indikator*; Report OR.16.18; Østfoldresearch: Fredrikstad, Norway, 2018.
9. Listrat, A.; Lebret, B.; Louveau, I.; Astruc, T.; Bonnet, M.; Lefaucheur, L.; Picard, B.; Bugeon, J. How Muscle Structure and Composition Influence Meat and Flesh Quality. *Sci. World J.* **2016**, *2016*, 3182746. [CrossRef]
10. Pearce, K.L.; Rosenvold, K.; Andersen, H.J.; Hopkins, D.L. Water distribution and mobility in meat during the conversion of muscle to meat and ageing and the impacts on fresh meat quality attributes—A review. *Meat Sci.* **2011**, *89*, 111–124. [CrossRef] [PubMed]
11. Hughes, J.M.; Oiseth, S.K.; Purslow, P.P.; Warner, R.D. A structural approach to understanding the interactions between colour, water-holding capacity and tenderness. *Meat Sci.* **2014**, *98*, 520–532. [CrossRef] [PubMed]

12. Orkusz, A. Effects of packaging conditions on some functional and sensory attributes of goose meat. *Poult. Sci.* **2018**, *97*, 2988–2993. [CrossRef]
13. McMillin, K.W. Where is MAP Going? A review and future potential of modified atmosphere packaging for meat. *Meat Sci.* **2008**, *80*, 43–65. [CrossRef]
14. Cheng, Q.; Sun, D.W. Factors affecting the water holding capacity of red meat products: A review of recent research advances. *Crit. Rev. Food Sci. Nutr.* **2008**, *48*, 137–159. [CrossRef]
15. Payne, S.R.; Durham, C.J.; Scott, S.M.; Devine, C.E. The effects of non-vacuum packaging systems on drip loss from chilled beef. *Meat Sci.* **1998**, *49*, 277–287. [CrossRef]
16. Jin, S.-K.; Kim, I.-S.; Song, Y.-M.; Kim, D.-H.; Lee, C.-Y.; Hur, I.-C.; Park, J.; Kang, S.-N.; Hur, S.-J. Effect of Packaging Methods on Quality Characteristics of Low-Grade Beef during Aging at 16C. *J. Food Process. Preserv.* **2013**, *37*, 1111–1118. [CrossRef]
17. Stella, S.; Bernardi, C.; Tirloni, E. Influence of skin packaging on raw beef quality: A review. *J. Food Qual.* **2018**, *2018*, 7464578. [CrossRef]
18. Droval, A.A.; Benassi, V.T.; Rossa, A.; Prudencio, S.H.; Paião, F.G.; Shimokomaki, M. Consumer attitudes and preferences regarding pale, soft, and exudative broiler breast meat. *J. Appl. Poult. Res* **2012**, *21*, 502–507. [CrossRef]
19. European Commission regulation (EC) No 450/2009 of 29 May 2009 on Active and Intelligent Materials and Articles Intended to Come into Contact with Food. Available online: https://eur-lex.europa.eu/LexUriServ/LexUriServ.do?uri=OJ:L:2009:135:0003:0011:EN:PDF (accessed on 2 May 2021).
20. Renerre, M. Factors involved in the discoloration of beef meat. *Int. J. Food Sci. Technol.* **1990**, *25*, 613–630. [CrossRef]
21. Dalgaard, P.; Gram, L.; Huss, H.H. Spoilage and shelf-life of cod fillets packed in vacuum or modified atmospheres. *Int. J. Food Microbiol.* **1993**, *19*, 283–294. [CrossRef]
22. Randell, K.; Hattula, T.; SkyttÄ, E.; Sivertsvik, M.; Bergslien, H.; Ahvenainen, R. Quality of filleted salmon in various retail packages. *J. Food Qual.* **1999**, *22*, 483–497. [CrossRef]
23. Jakobsen, M.; Bertelsen, G. The use of CO_2 in packaging of fresh red meats and its effect on chemical quality changes in the meat: A review. *J. Muscle Foods* **2002**, *13*, 143–168. [CrossRef]
24. Pettersen, M.K.; Hansen, A.Å.; Mielnik, M. Effect of Different Packaging Methods on Quality and Shelf Life of Fresh Reindeer Meat. *Packag. Technol. Sci.* **2014**, *27*, 987–997. [CrossRef]
25. Holck, A.L.; Pettersen, M.K.; Moen, M.H.; Sørheim, O. Prolonged shelf life and reduced drip loss of chicken filets by the use of carbon dioxide emitters and modified atmosphere packaging. *J. Food Prot.* **2014**, *77*, 1133–1141. [CrossRef]
26. Rotabakk, B.T.; Birkeland, S.; Jeksrud, W.K.; Sivertsvik, M. Effect of Modified Atmosphere Packaging and Soluble Gas Stabilization on the Shelf Life of Skinless Chicken Breast Fillets. *J. Food Sci.* **2006**, *71*, S124–S131. [CrossRef]
27. Lawless, H.T.; Heymann, H. *Sensory Evaluation of Food—Principles and Practices*; Springer: Berlin, Germany, 2010.
28. International Organization for Standardization. *ISO 13229 Sensory Analysis—General Guidance for Establishing a Sensory Profile*; International Organization for Standardization: Geneva, Switzerland, 2016.
29. International Organization for Standardization. *ISO 8589 Sensory Analysis—General Guidance for the Design of Test Rooms*; International Organization for Standardization: Geneva, Switzerland, 2007.
30. International Organization for Standardization. *ISO 8586 Sensory Analysis—General Guidance for Selection, Training and Monitoring of Selected Sensory Assessors and Expert Assessors*; International Organization for Standardization: Geneva, Switzerland, 2012.
31. RCore Team. *R: A Language and Environment for Statistical Computing*; R Foundation for Statistical Computing: Vienna, Austria, 2019.
32. Liland, K.H. mxml: Mixed Model ANOVA and Statistics for Education; Package Version 1.2.4. Available online: https://mran.microsoft.com/snapshot/2020-04-17/web/packages/mixlm/index.html (accessed on 2 May 2021).
33. Geesink, G.; Robertson, J.; Ball, A. The effect of retail packaging method on objective and consumer assessment of beef quality traits. *Meat Sci.* **2015**, *104*, 85–89. [CrossRef]
34. Rossaint, S.; Klausmann, S.; Kreyenschmidt, J. Effect of high-oxygen and oxygen-free modified atmosphere packaging on the spoilage process of poultry breast fillets. *Poult. Sci.* **2015**, *94*, 96–103. [CrossRef]
35. Fu, Q.-Q.; Liu, R.; Zhang, W.-G.; Li, Y.-P.; Wang, J.; Zhou, G.-H. Effects of Different Packaging Systems on Beef Tenderness Through Protein Modifications. *Food Bioprocess Technol.* **2015**, *8*, 580–588. [CrossRef]
36. Moczkowska, M.; Półtorak, A.; Montowska, M.; Pospiech, E.; Wierzbicka, A. The effect of the packaging system and storage time on myofibrillar protein degradation and oxidation process in relation to beef tenderness. *Meat Sci.* **2017**, *130*, 7–15. [CrossRef] [PubMed]
37. Lagerstedt, Å.; Lundström, K.; Lindahl, G. Influence of vacuum or high-oxygen modified atmosphere packaging on quality of beef M. longissimus dorsi steaks after different ageing times. *Meat Sci.* **2011**, *87*, 101–106. [CrossRef] [PubMed]
38. Zakrys-Waliwander, P.I.; O'Sullivan, M.G.; O'Neill, E.E.; Kerry, J.P. The effects of high oxygen modified atmosphere packaging on protein oxidation of bovine M. longissimus dorsi muscle during chilled storage. *Food Chem.* **2012**, *131*, 527–532. [CrossRef]
39. Lund, M.N.; Lametsch, R.; Hviid, M.S.; Jensen, O.N.; Skibsted, L.H. High-oxygen packaging atmosphere influences protein oxidation and tenderness of porcine longissimus dorsi during chill storage. *Meat Sci.* **2007**, *77*, 295–303. [CrossRef]
40. Herbert, U.; Kreyenschmidt, J. Comparison of Oxygen- and Nitrogen-Enriched Atmospheres on the Growth of Listeria Monocytogenes Inoculated on Poultry Breast Fillets. *J. Food Saf.* **2015**, *35*, 533–543. [CrossRef]

41. Pettersen, M.K.; Nissen, H.; Eie, T.; Nilsson, A. Effect of packaging materials and storage conditions on bacterial growth, off-odour, pH and colour in chicken breast fillets. *Packag. Technol. Sci.* **2004**, *17*, 165–174. [CrossRef]
42. Löwenadler, J.; Rönner, U. Determination of dissolved carbon dioxide by coulometric titration in modified atmosphere systems. *Lett. Appl. Microbiol.* **1994**, *18*, 285–288. [CrossRef]
43. Killeffer, D.H. Carbon dioxide preservation of meat and fish. *Ind. Eng. Chem.* **1930**, *22*, 140–143. [CrossRef]
44. Stansby, M.E.; Griffiths, F.P. Carbon dioxide in handling fresh fish—Haddock. *Ind. Eng. Chem.* **1935**, *27*, 1452–1458. [CrossRef]

Article

Application of PET/Sepiolite Nanocomposite Trays to Improve Food Quality

Teresa Fernández-Menéndez [1], David García-López [2], Antonio Argüelles [3], Ana Fernández [4] and Jaime Viña [1,*]

1 Department of Materials Science and Metallurgical Engineering, University of Oviedo, 33203 Gijón, Spain; tere_asturias@yahoo.es
2 SMRC Automotive Interiors Spain S.L.U., 47800 Medina de Rioseco, Spain; dgarci12@smrc-automotive.com
3 Department of Construction and Manufacturing Engineering, University of Oviedo, 33203 Gijón, Spain; antonio@uniovi.es
4 Klöckner Pentaplast, 33128 Vegafriosa, Spain; ana.fernandez@kpfilms.com
* Correspondence: jaure@uniovi.es

Abstract: New PET and nanosepiolite materials are produced for its application in innovative packaging with better performance. In our previous work, we demonstrate that the use of different percentages of sepiolite modified with different organosilanes improved mechanical and barrier properties of PET. Nanocomposites permeability can decrease up to 30% compared to that of pure PET and the mechanical analyses show that, although PET nanocomposites are more brittle than virgin PET, they are also harder. In the present work, we are going to study the properties of this innovative packaging with real food analyzing mechanical properties related to the product transport together with permeability and microbiological characteristics. At the same time, it has been seen that it is possible to lighten trays, which is very important both industrially and environmentally. On the other hand, a good quality packaging for food needs to ensure that organoleptic and physico-chemical characteristics of the product inside are not modified due to migration of any of the packaging material to the food itself. Results obtained in this work also show lower count of aerobic mesophilic bacteria and *Enterobacteriaceae* (EB), reducing the incidence of food contaminations by microorganisms.

Keywords: PET; sepiolite; nanocomposites; MAP; microbiological quality; chicken

Citation: Fernández-Menéndez, T.; García-López, D.; Argüelles, A.; Fernández, A.; Viña, J. Application of PET/Sepiolite Nanocomposite Trays to Improve Food Quality. *Foods* **2021**, *10*, 1188. https://doi.org/10.3390/foods10061188

Academic Editors: Valeria Rizzo, Muratore Giuseppe and Benu P. Adhikari

Received: 15 April 2021
Accepted: 21 May 2021
Published: 25 May 2021

1. Introduction

Food packaging has many useful functions, such as food containment, marketing, protection, and preservation during the shelf life of a product. In order to accomplish all this, a food packaging material must have enough strength to overcome its filling process, transport, and customer handling. At the same time, it needs to have the appropriate barrier properties for certain applications, such as in modified atmosphere packaging (MAP), and, of course, it needs to keep migration of packaging components to food to a minimum, complying with all regulation regarding Food Contact Materials (FCM), such as Food and Drug Administration (FDA) in USA, or European Regulation (EU) No 10/2011 of 14 January 2011.

Due to its low weight and versatility, among other things, polymers have been one of the most important materials used in food packaging. Plastic packaging plays a key role in protecting food from exterior virus and microorganisms as well as helping extend shelf life of packed food. However, it is also very important to keep migration of plastic components into food to a minimum, since that could ruin the product packed. Migration tests are regulated by legislation; thus, the amount of migrants allowed in the food is specified by the global migration limits. At the same time, the legislation records the lists of permitted substances to be used in food contact materials, this is the "positive list". Each of this substance has its migration limit specified, in order to avoid food toxicity. The analysis done for controlling the amount of each substance in the packaging are called specific migration tests.

Fresh products (poultry, fruits, vegetables, etc.) are a vehicle for the transmission of bacterial, parasitic, and viral pathogens capable of causing human illness [1]. Within food packaging applications, improving shelf life of packed poultry is a huge challenge for the industry. Fresh poultry meat is highly popular among consumers and, at the same time, it is highly perishable (rapid microbial growth) leading to high economic losses [2,3]. Its shelf life depends mainly on poultry handling and processing (in the initial number of microorganisms) [4] or on its storage conditions through all the food chain.

One of the most important food packaging systems is MAP, where the food is packed, together with a certain mixture of gases that will keep freshness of the food, enhancing preservation, and extending shelf life. MPA requires a tray or base container and a lid to seal the packaged content. There are many plastic materials that could be used for this application, being poly(ethylene terephthalate) (PET) one of the most widely used. Depending on the amount of barrier required for the application, sometimes it is needed to apply a multilayer film with a high barrier polymer in it. In the present work, the base material used for the tray will be PET as well as for the lid. Adding the nanoclay to the PET it is expected to obtain trays with improve mechanical and barrier properties. One of the most used clays when talking about nanocomposites is montmorillonite (MMT). However, the clay chosen for the work has been sepiolite, which is a magnesium silicate with the following formula: $Mg_8Si_{12}O_{30}(OH)_2(H_2O)_4\ 8H_2O$. It is a fibrous clay with nanometric dimensions that vary between 0.2 and 3 μm in length, 10–30 nm in width, and 5–10 nm in thickness, which gives the sepiolite a high aspect ratio of about 27. In addition, sepiolite has a surface area of about 300 $m^2\ g^{-1}$, and an outer layer of silanol groups. All these characteristics make the sepiolite perfect for surface modification with organosilanes and other reagents on its surface. It has also shown better mechanical properties than MMT in previous works [5].

The migration analysis of the trays, as well as microbiological tests are done to prove the possibility of using this material for food packaging, complying with actual legislation.

2. Materials

PET pellets, kindly supplied by LINPAC Packaging (Pravia, Spain), were from Novapet S.A. (Zaragoza, Spain). The pellets' intrinsic viscosity (IV) was 0.81 dL/g. PET-EVOH-PE laminated sheet is a PET sheet laminated with an EVOH-PE flexible film (its structure being EVA/PA/EVOH/PA/PE). The sealing top film of the tray is a film coated with aluminum oxide (AlOx) (BOPET), Mylar® 850 from DuPont Teijin Films UK Limited, Middlesbrough, England. Mylar® 850 is a co-extruded, one side amorphous, heat sealable polyester film, suitable for use in contact with food. Its oxygen transmission rate is 56 $cm^3/m^2/day$, at 23 °C and 60/70% RH, for a 30 micron film. Two types of sepiolite were supplied by Tolsa S.A. (Madrid, Spain), one modified with 2% of 3-metracyloxypropil trimetoxysilane (MEMO, CAS 2530-85-2) and the other one with 2% of 3-aminopropyltriethoxysilane (AMEO, CAS 919-30-2). Both organomodifiers are suitable for food packaging with restrictions regarding the amount of absorbed substance by kg of product packed, as stated in Regulation CE 975/2009 for MEMO (0.05 mg by kg of packed product) and in Directive 2007/19/CE for AMEO (between 0.05 mg to 3 mg by kg of product in the package). The nanocomposite masters were produced at Repol S.L. (Almazora Castellón, Spain) in an industrial polyamide low shear extruder.

3. Experimental Part

In order to produce nanocomposites at industrial level, it was necessary to do the first steps at laboratory scale, as shown in previous works [6]. The materials used in this paper, are those found to be the best ones in terms of processability and mechanical properties.

The first step of nanocomposite fabrication at industrial level was to produce the masterbatches.

The PET/nanosepiolite masters were produced at Repol S.L. facilities. Conditions in the production plant were optimized to minimize PET matrix degradation, reducing humidity, and decreasing extrusion shear on the nanocomposites. Two different masters with 10% sepiolite each were produced. In one master, the sepiolite was previously modified with MEMO, and the other one with AMEO in Tolsa S.A. Then, these masters were diluted to the final percentage of sepiolite into the extruder in order to obtain the corresponding sheets for characterization. The use of these modifiers reduces the sepiolite–sepiolite interactions, which favors a better dispersion, and alignment of nanofibers that translate in effectiveness in mechanical properties [7].

Drying conditions for the master were 80 °C for 7 h, and 120 °C for 7 h for the pristine PET. The drier used was a CRAMER-TROCKNER model PK 100/300F. With the aim of simplification, from now on virgin PET will be referred to as PET.

The industrial extruder used was a Luigi Bandera SpA twin screw extruder, from LINPAC Packaging. In this extruder it was obtained the nanocomposite's sheet that then is taken to a KIEFEL GmbH thermoforming machine to obtain the desired final trays. The trays chosen for this project are MAP trays, with the following dimensions: 18 cm width, 25 cm length, and 45 mm depth. In this work it will be referred to as B1825-45 tray. The nanocomposite trays are sent to a poultry packer (Sada, Nutreco. Spain). There, 2 kg of breast chicken are packed in a modified atmosphere containing 70% CO_2, 20% O_2, and 10% N_2 in each tray. Control trays are packed in the same way, in a PET tray. Then, samples are taken for microbiological analysis of mesophilic aerobes and *Enterobacteriaceae* for 14 days. In this work, the microbiological quality of chicken fillets was assessed by determining the number of mesophilic aerobic bacteria, and *Enterobacteriaceae*. These analyses will help us determine if microbial load of those species in chicken, packed in nanocomposite trays, is lower than that packed in regular PET, in order to have an idea of food sanitary quality.

3.1. TGA

Thermogravimetric Analysis (TGA) was used to determine nanosepiolite percentage within the nanocomposite sheets. The analyses were performed in a Mettler Toledo 851e equipment, using a procedure in two steps:

First step: from 50 to 600 °C at 20 °C/min under nitrogen atmosphere.

Second step: from 600 to 900 °C at 20 °C/min under air atmosphere.

3.2. Permeability

The permeability analyses were done on sheet samples; specimens were taken from the extruded sheets before going to thermoforming into trays.

Oxygen transmission rate (OTR) was measured in an OXTRAN with a volumetric sensor (Oxtran SS 2/20, MOCON. Barcelona, Spain). Previously to the analysis the samples were conditioned, 48 h under an atmosphere with 0% RH. Oxygen transmission rate was measured at 23 °C and 0% RH following Standard ASTM D3985-17 and the effective area exposed to permeation was 50 cm^2.

3.3. Puncture Test

Plastic products are more prone to fail when submitted to a multiaxial impact, rather than to a slow-motion load. In many applications, packaging materials are exposed to penetrating damages, which lead to barrier properties and package integrity deterioration. Thus, it is very important to obtain packaging materials with good impact strength properties that help preserve the food until its use.

These impact tests were done in an MTS-831 equipment, following ISO 6603-2:2000 methodology [8]. The speed used was 4.4 m/s and tests were at room temperature (23 °C).

The specimen, with an effective diameter of 40 mm, is hold with two anchor rings, then the impactor ($\varphi_{impactor}$ = 20 mm) hits on the specimen center from below. The curve strength versus strain is registered for each sample, together with absorbed energy (E). However, it is very important to describe the failure mode in order to know if the material is going to break in a fragile way, a ductile, or in any of the intermediate modes in between [8]. In a ductile break (D), the specimen breaks slowly deforming the material with the absorbed E, while the additional, non-absorbed E, is used to extend the crease (Dc). On the contrary, in a fragile break (F) the crease is spread quickly, suddenly, and totally, causing the break of the sample.

3.4. Compression Test

Lateral compression tests and stiffness were tested on a Hounsfield H1KS Benchtop equipment following LINPAC Packaging internal procedures for trays. For each trial, between 65 and 70 trays were tested.

3.5. Microbiological Tests

Twelve specimens of the nanocomposite trays were analyzed for each material, and sixteen for control.

For total viable count (TVC) determination, 25 g of superficial meat are taken aseptically. Samples are mixed with 225 mL of buffered peptone water and is then homogenized in a Stomacher® (dilution 1:10). After that, 1 mL sample is taken from the main dilution and then the dilutions needed to obtain an appropriate number of microorganisms are done. The incubation time and temperature for mesophilic aerobes are 72 h at 30 °C and a Petrifilm Aerobic Count Plate is used (ISO 4833-1, 2003 [9]), and 24 h at 37 °C for Enterobateriacae using a 3M Petrilm to help counting (ISO 21528-2, 2004 [10]). Bacterial count results are expressed in $\log_{10}$ of colony-forming units per gram of meat (log cfu/g).

Microbiological analyses were done, on 3 samples per day and treatment, on the following days post packaging: 2, 7, 10, and 14. At the same time, head space gases were measured to see the evolution during the microbial study, using an OXIBABY-M O2/CO2 (WITT-GASETECHNIK. Witten, Germany). The specimens were kept at 5 °C during all that period.

4. Results and Discussion

It has been analyzed before [6], the effect caused by different nanosepiolite masterbatches concentration (one with 20%, and the other one with 10%) on the final nanocomposite properties. It was concluded that those nanocomposites coming from a less concentrated master had better homogeneity, as well as viscosity and mechanical properties of the final material. Thus, for this study, a 10% nanosepiolite master was aimed. This percentage of sepiolite is theoretical because after dosing the sepiolite in an industrial extruder, the final amount of nanoclay in the nanocomposite changes. This is due to the difficulty of adding a powder to an industrial extruder at 600 kg/h. In this way, a PET masterbatch with 7.66% sepiolite modified with MEMO (MB1), and another one with 8.56% of sepiolite modified with AMEO (MB2) were used in this study determined by TGA.

Table 1 shows the percentages of sepiolite in the nanocomposites, after dilution of the masters into the PET industrial extruder.

Table 1. Samples and sepiolite content on the nanocomposites.

Master	Sample	Nanocomposite	TGA (% nS)
-	M0	Virgin PET	0
MB1	M1	PET + 1.20% nS_MEMO	1.20
	M2	PET + 1.37% nS_MEMO	1.37
	M3	PET + 1.78% nS_MEMO	1.78
MB2	M4	PET + 1.58% nS_ AMEO	1.59
	M5	PET + 1.65% nS_ AMEO	1.65
	M6	PET + 1.88% nS_ AMEO	1.88

4.1. Permeability

The permeability was analyzed on industrially extruded sheet samples with different thickness. Permeability to oxygen, calculated as OTR, improved in all nanocomposite's samples compared to that of pure PET. The improvement can be up to 30% with 1.37% of nanosepiolite. It is shown the OTR/Sheet Thickness, which is calculated dividing the permeability, given in OTR units, by the sample thickness. It was observed that increasing the clay concentration did not change the permeability of the samples substantially (Figure 1a). In addition, as stated by Ke and Yongping [11] and tested in this work, the processability of the nanocomposite is much more difficult when increasing the amount of nanosepiolite over 3% [6,12]. It is possible to decrease the amount of nanoclay in the samples, as long as the nanoparticles are well dispersed and oriented within the matrix [13–16]. However, if the sepiolite content is too low it will not do the job and if it is too much, PET matrix viscosity will decrease and it will not disperse properly, opening the path for gas and vapor molecules.

Comparing the nanocomposite samples to a PET-EVOH-PE laminated sheet, which is the sheet normally used when high barrier is required, we can see 20% improvement in the sample containing 1.37% of nanosepiolite modified with MEMO (sample M2). There are no mayor differences in terms of permeability performance between the two silanes used in this study.

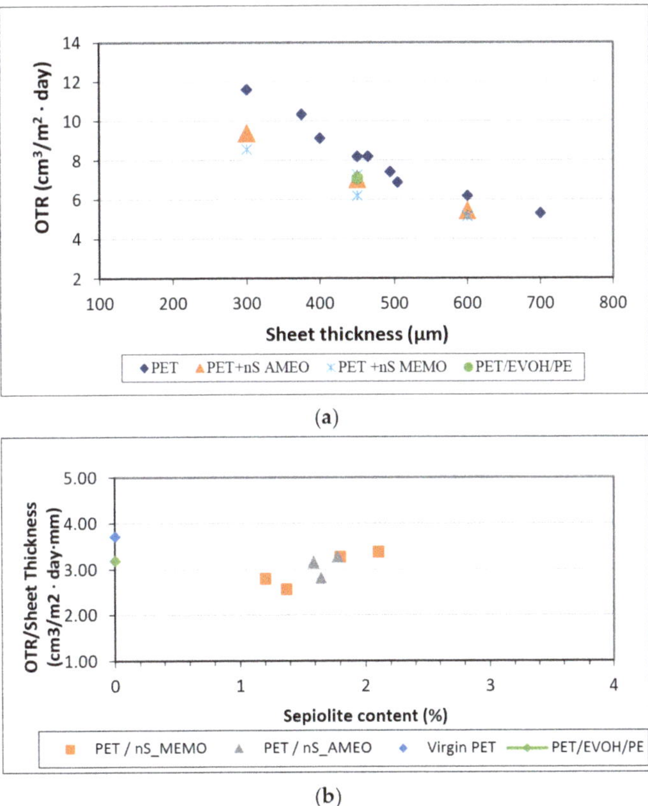

Figure 1. Oxigen transmission rate (OTR)/Sheet Thickness as a function of sepiolite content on the nanocomposite samples (**a**); OTR for nanocomposites as a function of sheet thickness (**b**).

Looking at Figure 1b it can be seen that, in order to achieve the same permeability of a 450 µm PET sample, a nanoPET composites could decreased its thickness in about 100–150 µm. These results are really important in terms of reducing PET consumption, which is good for both company savings in materials and transport, and for the environment.

4.2. Mechanical Properties of Nanocomposite Trays

In order to overcome its transport from factory to houses, including customer's handling, trays must be tough enough. That is why it is so important to test the packages with tests that can simulate its treatment once at the market.

These tests are done on the industrially extruded sheet before the thermoforming process. Registered curves for impact tests show a maximum which is related to the initial damage on the sheet, corresponding to the starting point of the fissure that will develop in a fracture. Analyzing the curves obtained in this test, maximum load and its associated deformation can be known, as well as perforation energy. On the other hand, this test shows the way the sample breaks allowing us to define the failure mode of each specimen.

Thickness is measure on extruded sheet, and the results shown in Table 2 are the average obtained in all the samples width (795 mm). Since this measure is of great importance in impact results, we will compare samples with the same thickness. Thus, for a sheet with 465 µm, a sample M5 with 1.65% nS shows an 8% improvement in impact strength [14]; nevertheless, the impact energy decreases 18% in the nanocomposite compared to that of pure PET (with 467 µm) (Table 2). This means there is less energy for deformation, but it takes a little higher force to break the nanocomposite sheet. The same behavior is seen in nanocomposites with MEMO, but impact and energy values are a little lower than in those modified with AMEO.

Table 2. Impact puncture results for nanocomposites containing nanosepiolite modified with MEMO (3-metracyloxypropil trimetoxysilane) and with AMEO (3-aminopropyltriethoxysilane).

Sample	TGA (% nS)	Thickness (µm)	Max Load (N)	Puncture Energy (J)	Failure Mode *	Picture
			MEMO			
M0	0	467 ± 2	720 ± 16	6.6 ± 0.8	D	
M2	1.37	351 ± 12	560 ± 42	2.6 ± 0.6	FD	
M3	1.78	470 ± 11	730 ± 13	5.0 ± 2.4	FD-D	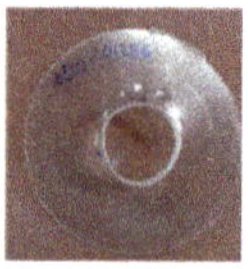
M1	1.20	595 ± 15	1000 ± 23	8.3 ± 0.9	D	

Table 2. *Cont.*

Sample	TGA (% nS)	Thickness (μm)	Max Load (N)	Puncture Energy (J)	Failure Mode *	Picture
			AMEO			
M0	0	467 ± 2	720 ± 16	6.6 ± 0.8	D	
M4	1.59	320 ± 6	520 ± 15	3.1 ± 0.7	FD-Dc	
M5	1.65	465 ± 9	780 ± 34	5.4 ± 0.7	Dc	
M6	1.88	601 ± 12	1100 ± 40	8.5 ± 0.4	D	

* Failure Modes: D—Ductile, FD—transition Fragile/Ductile, Dc—Ductile with crease.

A PET sheet (100% virgin) shows a ductile behavior, whilst nanocomposite materials evidence a fragile component when using MEMO modified sepiolite (see Table 2). However, samples modified with AMEO do not show fragile behavior but a ductile with a forming crease one. Otherwise, there are no notable differences in nanocomposites impact behavior depending on the organomodifier used.

Regarding sample's thickness, when increasing this value, maximum load, impact energy, and ductility at break increases in all the samples no matter the sepiolite content or its organic modifier.

In order to normalize result to thickness, data in Newtons were divided by the thickness of each sample. The results show that samples modified with AMEO improve its maximum load at impact when increasing nanosepiolite content, whereas samples with MEMO have poorer results. Impact load increases in all samples with AMEO over that of pure PET, with a 19% increase in sample containing 1.88% nanosepiolite (M6). Meanwhile, the maximum load improvement in MEMO samples is of 9% over that of virgin PET.

4.3. Compression Tests

Compression tests results are done with the tray in vertical position since is, generally, the most critical force MAP trays are going to be subjected to. Results indicate that the nanocomposite trays have more resistance to compression forces compare to the pure PET trays.

Nanocomposites are more resistant to lateral compression than pure PET. The improvement can be 66.7% for the sample M6 containing 1.88% nS.

In the following graph (Figure 2) it is shown the stiffness per gram. The results show that for the same tray weight the stiffness increase can be up to 85% with 1.88% nanosepiolite AMEO modified. Thus, the addition of nanosepiolite to a PET matrix results in increased stiffness and decreased ductility, as seen in these results [6,15]. This could be due to a good interaction between the PET matrix and the sepiolite. The high aspect ratio of this clay is key in this interaction, together with the organomodifiers used [16,17]. It is also seen here an increase in stiffness with the amount of sepiolite when using AMEO as organomodifier, an increase of 0.23 g nanosepiolite produces an improvement of 40% in stiffness. Whereas in a sample with nearly double sepiolite content modified with MEMO (1.09 g more), there is an increase of just 4%.

It is seen that nanocomposites sheets with around 1.6% nanosepiolite need 150 μm less thickness in order to generate the same stiffness as a pure PET samples. Thus, a sample with 1.65% nanosepiolite modified with AMEO (M5) has the same stiffness than a sample that weights 8 g more in pure PET.

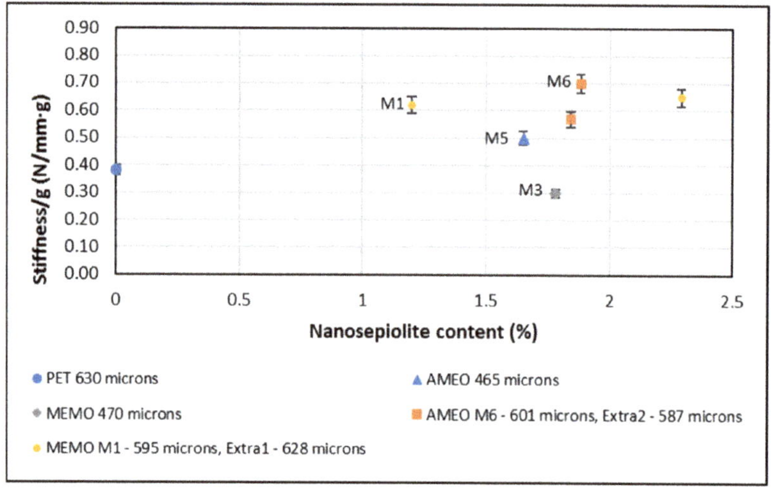

Figure 2. Nanocomposites stiffness per gram compared to that of pure PET.

As it is shown in this paper, results with both AMEO and MEMO as organic modifiers for the sepiolite have quite similar performance. However, those nanocomposites with AMEO are more resistant to lateral compression forces, a little stronger and less brittle. Moreover, between the two nanocomposite's types there are some other differences which are important for good industrial extrusion and thermoforming processes. These differences are mainly that AMEO nanosepiolite seems to disperse better into the PET matrix [6], and its processability in an industrial extruder was a little better, showing a more stable behavior.

As a function of the results obtained, the sheet chosen for the next step is that containing between 1.5 and 1.9% of nanosepiolite modified with 2% of AMEO. At the same time, it will be used the trays with intermediate thickness. Thus, the sheet sample chosen to thermoformed trays to be used in the next step is M5. These trays were thermoformed in order to do microbiological analysis and migration test studies.

4.4. *Microbial Analysis in Packed Chicken Breast*

Microbiological analysis is done on breast chicken packed in the same trays as discussed previously, a B1825-45 tray made with M5 sheet (Tray A1, 465 m) and using a virgin PET tray in 601 μm as control (Tray A0). This thickness is the regular one that would be used for this type of product. The tray thickness refers to the average thickness of the sheet before thermoforming (the average value of all sheet width).

Mesophilic aerobes and *Enterobacteriaceae* family include not only pathogenic species, but environmental species as well, which often appear in the food manufacturing environment without posing any health hazard. In fresh food, high number counts are not recommended, although an elevated count does not imply the presence of pathogenic flora. However, the total count reflects sanitary quality of the analyzed products. If the total count is high, testing of specific pathogens can be done. At the same time, a low count does not mean the sample is pathogen free, it depends on the composition of the microbiota [18]. This control should be done in accordance with Commission Regulation (EC) No 2073/2005.

In the following graphs (Figure 3) it is shown microbiology parameter's growth, together with atmosphere gases evolution with packing time for both samples chosen; which is to say, 2 kg of breast chicken packed in trays A1, and also the control samples which are packed in pure PET (A0). In both cases, the top film is BOPET coated with AlOx. Results shown in the graphics correspond to the average count obtain for the three samples examined, for each period (2, 7, 10, and 14 days).

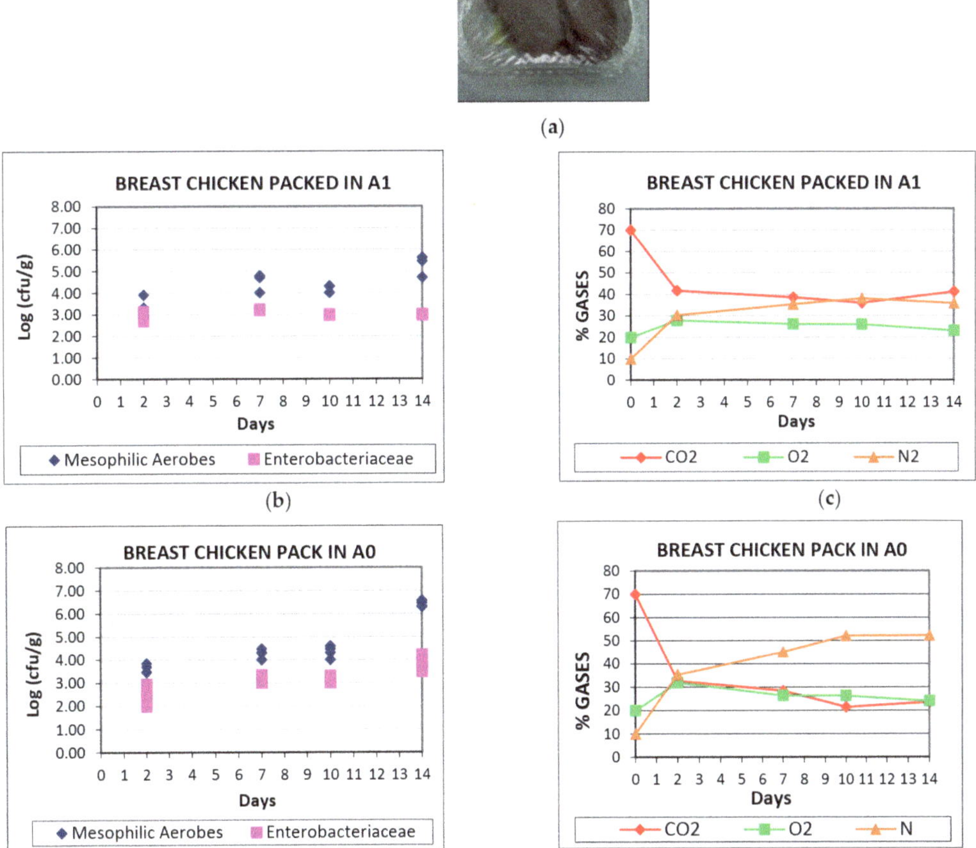

Figure 3. Picture of the packed chicken (**a**), aerobic mesophilic and *Enterobacteriaceae* count (**b**,**d**) and gases concentration evolution in 2 kg chicken breast (**c**,**e**), packed in A1 and A0 nanocomposite tray, respectively.

Generally, in hygienically handled fresh food, with a limited shelf life, if the count of microorganism is high it will decrease the serviceable time of the products. The chicken microbiological quality was initially good in both trays, being at day 2 (first count analyses) lower than 4 log (cfu/g) [19,20]. At day fourteenth, none of the samples had arrived at 7 log (cfu/g) which is the maximum limit stablished for fresh chicken [21–23]. Looking at mesophilic aerobes results (Figure 3a,d), it can be seen the values of the two trays (A0, A1) are quite similar up until the fourteenth day, where results on nanocomposite's tray show lower numbers of colony forming units of mesophilic aerobes in those trays made with nanocomposite, up to one order of magnitude lower (from log (cfu/g) 6.57 to 5.25).

Regarding total count of *Enterobacteriaceae* (EB), the average value of PET trays analyzed is 3.83 log (cfu/g) (8.25×10^3 ufc/g) whilst this number is 2.98 log (cfu/g) (9.67×10^2 ufc/g) for the nanocomposite tray. Thus, nanosepiolite trays have lower microbiological charge. EB remains practically constant and very low from day 2 to 14 for the nanocomposite trays, starting at an average log (cfu/g) of 2.83 and ending, at 14 days, at 2.98, whereas in PET trays, although the starting count of *Enterobacteriaceae* is lower, log (cfu/g) 2.43, at day 14 it is of 3.80. The explanation for the lower growth of bacterial counts could be due not only to CO_2 [24] but also to N_2 capability of inhibiting bacterial growth [25,26].

Figure 3b,d shows head space gases evolution for the breast chicken packed in nanocomposite's trays and in PET tray. CO_2 content decrease is mainly due to product absorption into the food [21–28], but also to loss of the gas through permeation across the plastic. Major CO_2 absorption takes place during the first two days in the two samples, resulting in an equilibrium concentration, lower than the initial [29]. The absorption of CO_2 depends on partial pressure in the head space of the packed tray [30] and also on the product itself (for example, chicken with skin or without it). In this study, these factors are the same for the two trays mentioned. However, the final CO_2 content is approximately 20% higher in nanocomposite's trays. Thus, in spite of the lower thickness of the new trays, the decrease in permeability helps in keeping CO_2 inside the package.

Regarding O_2, this was maintained quite constant throughout the 14 days due to initial amount being very similar to atmosphere O_2 concentration (approximately 20.9%). O_2 percentage increased during the first two days, due to the high decrease of CO_2 in that period, which caused loss of volume, and to the lack of diffusion of O_2 to the exterior since its partial pressure is nearly the same in and outside the package. After those days, O_2 remained quite constant for the reason already mention. It can be also seen in the gas's evolution figures that N_2 expands at the same time that CO_2 is absorbed into the chicken.

The improved barrier performance of the PET/sepiolite nanocomposites has, therefore, produced a decreased on microbial counts in the poultry trays due to their ability to keep modified atmosphere gases longer [12,31,32].

4.5. Migration

Global migration test for food contact plastics are done in a variety of food simulants, depending on the food type it is aimed to contain (fatty food, vegetables, etc.) and the storage conditions. In this work, global migration tests have been done following specifications under UNE-EN 1186-14 using iso-octane as a fatty food simulant.

For these analyses, the samples chosen were M3 and M6 (PET + 1.78% nS_MEMO and PET + PET + 1.88% nS_AMEO, respectively), since these are the ones with the highest sepiolite content. Global migration limits within European Regulation (UE) No 10/2011 and RD 866/2008 are 10 mg/dm^2 or 60 mg/kg. Results on both trays are lower than 1.0 mg/dm^2; thus, both trays (M3 and M6) comply with global migration limits for the chosen simulant and under the test conditions.

Regarding specific migration limits for MEMO and AMEO (0.05 mg/dm^2), being the results 0.04 mg/dm^2 for both MEMO and AMEO, it can be concluded that both trays also comply with the limits stablished in Directive 2002/72/EC and its subsequent amendments (Regulation EC 975/2009) for this product types. Actual legislation for food contact materials does not contain specific migration limits for natural silicates (except asbestos) such as sepiolite. However, since sepiolite is a magnesium silicate, the global migration of silicon and magnesium in iso-octane has been analyzed in order to know if there is any sepiolite migration from tray material to the packaged food. Results for silicon migration are lower than 0.05 mg/kg in both trays (M3 and M6), and results for magnesium are lower than 1.0 mg/kg. For this reason, it is concluded that migration of these two elements to the packaged food is well under the legislation limit.

5. Conclusions

The results obtained in the study have showed that the global properties of the trays have been improved.

It is possible to improve PET permeability by 30% with a nanocomposite containing just 1.37% of sepiolite in its matrix. It would be possible to decrease tray thickness in about 100–150 μm in order to obtain the same permeability of a 450 μm PET sample. These results would imply company savings in terms of materials and transport, and also it would be great for the environment, using less raw materials.

Mechanical properties have been also measured. In the impact test, it has been seen an improvement in the maximum load at impact in all the samples but in M3, which value is equal to that of the PET. In those samples modified with AMEO, impact energy is the same as to a PET sample with 1.88% sepiolite, whilst for the MEMO modified samples the best one, nearly equal to that of PET in impact energy is that one with 1.2% clay.

Failure behavior is ductile for a 100% virgin PET and has a fragile component in the nanocomposite sheet with MEMO modified sepiolite. However, samples modified with AMEO does not show fragile behavior but a ductile with a forming crease one. Otherwise, MEMO and AMEO samples show similar mechanical performances.

Nanocomposites trays are stiffer than PET ones and show better performance to lateral compression forces. This type of compression is most critical for sealed trays, since the top film tends to produce tension and, if the tray is not stiff enough, it will bend or even collapse easily. This would give a poor packaging impression to customers. It has been seen that it is possible to lighten trays with the use of nanosepiolite clay, which is very important both industrially and environmentally.

The final nanocomposite trays show lower numbers of colony forming units of mesophilic aerobes in those trays made with nanocomposite, up to one order of magnitude lower (from log(cfu/g) 6.57 to 5.25). EB count remains practically constant and very stable from day 2 to 14 for the nanocomposite trays, being the final count one order of magnitude lower than that obtained in a virgin PET tray (from log(cfu/g) 3.83 to 2.98).

It is important to note that nanocomposite trays are 136 μm less thick than that of the PET serving as control. Thus, lighter nanosepiolite trays with lower microbiological charge can be produced at industrial machines. Moreover, these nanocomposite trays comply with European Regulation (UE) No 10/2011 and RD 866/2008 for materials intended to be in contact with food products.

Author Contributions: Conceptualization, D.G.-L. and A.F.; methodology, A.A.; validation, J.V.; investigation, T.F.-M.; resources, A.F.; writing-review and editing, T.F.-M.; supervision, J.V. All authors have read and agreed to the published version of the manuscript.

Funding: This research received no external funding.

Data Availability Statement: The data presented in this study are available on request from the corresponding author. The data are not publicly available due to privacy restrictions of the companies involved.

Acknowledgments: Authors of this paper would like to thank Klöckner Pentaplast in Pravia (former LINPAC Packaging S.A.U.), Cidaut, Ainia, Grupo Repol S.A. and Tolsa S.A. for their different collaborations in this project.

Conflicts of Interest: The authors declare no conflict of interest.

Nomenclature

PET	Poly(ethylene terephthalate)
Enterobacteriaceae	EB
MAP	Modified Atmosphere Packaging
FCM	Food Contact Materials
FDA	Food and Drug Administration
nS	Nanosepiolite
MMT	Montmorillonite
IV	Intrinsic Viscosity
PA	Polyamide
PE	Polyethylene
EVA	Ethylene Vinyl Acetate
EVOH	Ethylene Vinyl Alcohol
AlOx	Aluminum Oxide
BOPET	Bi-oriented Poly(ethylene terephthalate)
MEMO	3-methacryloxypropil trimethoxysilane
AMEO	3-aminopropyltriethoxysilane
TGA	Thermogravimetric Analysis
OTR	Oxygen Transmission Rate
TVC	Total Viable Counts
cfu	Colony forming units
Max	Maximum

References

1. Abadias, M.; Usal, J.; Anguera, M.; Solsona, C.; Viñas, I. Microbiological quality of fresh minimally processed fruit and vegetables, and sprouts from retail establishments. *Int. J. Food Microbiol.* **2008**, *123*, 121–129. [CrossRef]
2. Khanjari, A.; Karabagias, I.K.; Kontominasm, M.G. Combined effect of N,O-Carboxymethyl chitosan and oregano essential oil extended shelf life and control listeria monocytogenes in raw chicken meat fillets. *Food Sci. Tecnol.* **2013**, *53*, 94–99.
3. Pires, J.R.A.; de Souza, V.G.L.; Fernando, A.L. Chitosan/montmorillonite bionanocomposites incorporated with rosemary and ginger essential oil as packaging for fresh poultry meat. *Food Packag. Shelf* **2018**, *17*, 142–149. [CrossRef]
4. Yashoda, K.P.; Sachiondra, N.M.; Sakhare, P.Z.; Rao, D.N. Microbiological quality of broiler chicken carcasses processed hygienically in small-scale poultry processing unit. *J. Food Qual.* **2001**, *24*, 249–259. [CrossRef]
5. Gallego, R.; García-López, D.; Merino, J.C.; Pastor, J.M. How do shape of clay and type of modifier affect properties of polymer blends? *J. Appl. Polym. Sci.* **2013**, *127*, 3009–3016. [CrossRef]
6. Fernández-Menéndez, T.; García-López, D.; Argüelles, A.; Fernández, A.; Viña, J. Industrially produced PET nanocomposites with enhanced properties for food packaging applications. *Polym. Test.* **2020**, *90*, 106729. [CrossRef]
7. García-López, D.; Fernández, J.F.; Merino, J.C.; Santarén, J.; Pastor, J.M. Effect of organic modification of sepiolite for PA 6 polymer/organoclay nanocomposites. *Compos. Sci. Technol.* **2010**, *70*, 1429–1436. [CrossRef]
8. ISO. *Plastics-Determination of Puncture Impact Behavior of Rigid Plastics. Part 2: Instrumented Impact Testing*; ISO 6603-2:2000; ISO: Geneva, Switzerland, 2000.
9. ISO. *Microbiology of the Food Chain-Horizontal Method for the Enumeration of Microorganisms. Part 1: Colony Count at 30 Degrees C by the Pour Plate Technique*; ISO 4833-1:2013; ISO: Geneva, Switzerland, 2003.
10. ISO. *Microbiology of Food and Animal Feeding Stuffs-Horizontal Method for the Detection and Enumeration of Enterobateriaceae-Part 2: Colony-Count Method*; ISO 21528-2:2004; ISO: Geneva, Switzerland, 2004.
11. Ke, Z.; Yongping, B. Improve the gas barrier property of PET film with montmorillonite by in situ interlayer polymerization. *Mater. Lett.* **2005**, *59*, 3348–3351. [CrossRef]
12. Frounchi, M.; Dorubash, A. Oxygen barrier properties of poly (ethylene terephthalate) nanocomposite films. *Macromol. Mater. Eng.* **2009**, *294*, 68–74. [CrossRef]
13. Nielsen, L.E. Models for the permeability of filled polymer systems. *J. Macromol. Sci. A* **1967**, *1*, 929–942. [CrossRef]
14. Bharadwaj, R.K. Modeling the barrier properties of polymer-layer silicate nanocomposites. *Macromolecules* **2001**, *34*, 9189–9192. [CrossRef]
15. Cussler, E.L.; Hughes, S.E.; Ward, W.J.; Aris, R. Barrier membranes. *J. Membr. Sci.* **1988**, *38*, 161–174. [CrossRef]
16. Fredrikson, G.H.; Bicerano, J. Barrier properties of oriented disk composites. *J. Chem. Phys.* **1999**, *110*, 2181–2188. [CrossRef]
17. Tanaka, F.H.; Cruz, S.A.; Canto, L.B. Morphological, thermal and mechanical behaviour of sepiolite-based poly (ethylene terephthalate)/polyamide 66 blend nanocomposites. *Polym. Test.* **2018**, *72*, 298–307. [CrossRef]
18. Velasquez, E.; Garrido, L.; Valenzuela, X.; Galotto, M.J.; Guarda, A.; de Dicastillo, C.L. Physical properties and safety of 100% post-consumer PET bottle-organoclay nanocomposites towards a circular economy. *Sustain. Chem. Pharm.* **2020**, *17*, 100285. [CrossRef]

19. Valapa, R.B.; Loganathan, S.; Pugazhenthi, G.; Thomas, S.; Varghese, T.O. An overview of polymer-clay nanocomposites. In *Clay-Polymer Nanocomposites*; Jlassi, K., Chemeri, M.M., Thomas, S., Eds.; Elsevier: Amsterdam, The Netherlands, 2017; pp. 29–81.
20. Franke, C.; Höll, L.; Langowski, H.C.; Petermeier, H.; Vogel, R.F. Sensory evaluation of chicken breast packed in two different modified atmospheres. *Food Packag. Shelf* **2017**, *13*, 66–75. [CrossRef]
21. Latou, E.; Mexis, S.F.; Badeka, A.V.; Kontakos, S.; Kontominas, M.G. Combined effect of chitosan and modified atmosphere packaging for shelf life extension of chicken breast fillets. *LWT-Food Sci. Technol.* **2014**, *55*, 263–268. [CrossRef]
22. Chmiel, M.; Roszko, M.; Hac-Szymanczuk, E.; Adamczak, L.; Florowski, T.; Pietrzak, D.; Cegielka, A.; Bryta, M. Time evolution of microbiological quality and content of volatile compounds in chicken fillets packed using various techniques and stored under different conditions. *Poult. Sci.* **2020**, *6*, 1107–1111. [CrossRef]
23. Senter, S.D.; Arnold, J.W.; Chew, V. APC values and volatile compounds formed in commercially processed, raw chicken parts during storage at 4 and 13 °C under simulated temperature abuse conditions. *J. Sci. Food Agr.* **2000**, *80*, 1559–1564. [CrossRef]
24. Balamatsia, C.C.; Patsias, A.; Kontominas, M.G.; Savvaidis, I.N. Possible role of volatile amines as quality-indicating metabolites in modifies atmosphere-packaged chicken fillets: Correlation with microbiological and sensory attributes. *Food Chem.* **2007**, *104*, 1622–1628. [CrossRef]
25. Al-Nehlawi, A.; Saldo, J.; Vega, L.F.; Guri, S. Effect of high carbon dioxide atmosphere packaging and soluble gas stabilization pre-treatment on the shelf-life and quality of chicken drumsticks. *Meat Sci.* **2013**, *94*, 1–8. [CrossRef]
26. Munsch-Alatossava, P.; Gursoy, O.; Alatossava, T. Potential of nitrogen gas (N_2) to control psychrotrophs and mesophiles in raw milk. *Microbiol. Res.* **2010**, *165*, 122–132. [CrossRef] [PubMed]
27. Schumpe, A.; Quicker, G.; Deckwer, W.D. Gas solubilities in microbial culture media. *Adv. Biochem. Eng.* **1982**, *24*, 1–38.
28. Finne, G. Modified- and controlled-atmosphere storage of muscle foods. *Food Technol.* **1982**, *36*, 128–133.
29. Jiménez, S.M.; Salsi, M.S.; Tiburzi, M.C.; Rafaghelli, R.C.; Pirovani, M.E. Combined use of acetic acid treatment and modified atmosphere packaging for extending the shelf-life of chilled chicken breast portions. *J. Appl. Microbiol.* **1999**, *87*, 339–344. [CrossRef] [PubMed]
30. McMullen, L.M.; Stiles, M.E. Changes in microbial parameters and gas composition during modifies atmosphere storage of fresh pork loin cuts. *J. Food Protect.* **1991**, *54*, 778–783. [CrossRef]
31. Baker, R.C.; Hotchkiss, J.H.; Qureshi, R.A. Elevated carbon dioxide atmospheres for packaging poultry, I. Effects on ground chicken. *Poult. Sci.* **1985**, *64*, 328–332. [CrossRef]
32. Meredith, H.; Valdramidis, V.; Rotabakk, B.T.; Sivertsvik, M.; McDowell, D.; Bolton, D.J. Effect of different modified atmospheric packaging (MAP) gaseous combinations on campylobacter and shelf-life of chilled poultry fillets. *Food Microbiol.* **2014**, *40*, 196–203. [CrossRef]

Article

Improving the Storability of Cod Fish-Burgers According to the Zero-Waste Approach

Flavia Dilucia, Valentina Lacivita, Matteo Alessandro Del Nobile * and Amalia Conte

Department of Agricultural Sciences, Food and Environment, University of Foggia, Via Napoli, 25, 71121 Foggia, Italy; flavia.dilucia@unifg.it (F.D.); valentina.lacivita@unifg.it (V.L.); amalia.conte@unifg.it (A.C.)
* Correspondence: matteo.delnobile@unifg.it

Abstract: This research explored the potential of the zero-waste concept in relation to the storability of fresh food products. In particular, the prickly pear (*Opuntia ficus-indica*) peel (usually perceived as a by-product) and the pulp were dehydrated, reduced in powder, and used as food additives to slow down the growth of the main spoilage microorganisms of fresh cod fish burgers. The proportion between peel and pulp powder was such as to respect the zero-waste concept. The antibacterial activity of the peel and pulp in proper proportion was first assessed by means of an in vitro test against target microorganisms. Then, the active powder was added at three concentrations (i.e., 2.5 g, 7.5 g, and 12.5 g) to cod fish burgers to assess its effectiveness in slowing down the microbial and sensory quality decay of burgers stored at 4 °C. The results from the in vitro test showed that both the peel and pulp were effective in delaying microbial growth. The subsequent storability test substantially confirmed the in vitro test results. In fact, a significant reduction in growth rate of the main fish spoilage microorganisms (i.e., *Pseudomonas* spp., psychrotrophic bacteria, and psychrotolerant and heat-labile aerobic bacteria) was observed during 16 days of refrigerated storage. As expected, the antimicrobial effectiveness of powder increased as its concentration increased. Surprisingly, its addition did not affect the sensory quality of fish. Moreover, it was proven that this active powder can improve the fish sensory quality during the storage period.

Keywords: antimicrobial activity; fish storability; prickly pear cactus; by-products; sustainable approach; zero-waste

Citation: Dilucia, F.; Lacivita, V.; Nobile, M.A.D.; Conte, A. Improving the Storability of Cod Fish-Burgers According to the Zero-Waste Approach. *Foods* **2021**, *10*, 1972. https://doi.org/10.3390/foods10091972

Academic Editors: Valeria Rizzo and Muratore Giuseppe

Received: 14 July 2021
Accepted: 21 August 2021
Published: 24 August 2021

1. Introduction

Over the past few years, the amount of food waste produced and lost through the supply chain has become a severe problem for the world, causing nutrient loss, climate changes due to the production of greenhouse gas, and losses of resources like water and cultivated land [1].

One of the approaches suggested to reduce this global concern is the concept of zero-waste, a philosophy that prompts to find a way for all products to be recycled, so that no kind of waste will be sent to landfills or incinerators [2]. Large quantities of waste are generated during processing of fruit and vegetables. The by-products (seeds, peels, and pomace) represent about 25–30% of them [3]. These are discarded in landfill or incinerated, creating serious environmental complications and economic expenses, used for animal feeding or for the production of biogas or bio-fertilizers, but they possess many bioactive compounds, like phenolic acids, flavonoids, vitamins, as well as antioxidant and antimicrobial activity. As a consequence, these by-products can be applied to food fortification as a source of valuable bio-components, as well as for food packaging to enhance film performance [4,5].

Prickly pear cactus, *Opuntia ficus-indica* (L.) *Miller*, a Cactateae, is a tropical or subtropical plant, mostly deriving from Mexico, but also found throughout the American continent, over the Mediterranean basin and in southern Spain [6]. The world production of prickly pear is about 1 million tons per year [7]. Based on the cultivar and ripening stage, prickly pear consists of peel (accounts for 33 to 55%), pulp (accounts for 45 to 67%) and seeds (accounts for 2 to 10%) [7,8]. It is widely used as fresh fruit or for manufacturing fruit juice [9] and alcoholic beverages [10]. The peel, that represents the major waste in prickly pear, is considered an agricultural by-product, even though it is a source of dietary fibers, proteins and antioxidant compounds [7]. Additionally, both the fruit and the peel are rich in polyphenolic compounds that show biological and antimicrobial activity against different microorganisms [9,11–13]. A few applications of prickly pear by-products to foods are available. In particular, Palmeri et al. [12] used whole prickly pear extract to improve the shelf life of sliced beef. Chougui et al. [14] studied the application of the hydro-ethanolic extract of prickly pear peel to preserve margarine. In addition, some authors have used prickly pear peel powder as a functional ingredient to formulate bread and biscuits [7,15]. To the best of our knowledge, no zero-waste production has been proposed with prickly pears. However, giving the potential properties of this fruit, it can be supposed that the efforts in food science to recycle prickly pear by-products could gain industrial relevance.

Seafood products are dynamic and prone to innovation, with fish being highly appreciated but very often considered a time-consuming food to prepare. Therefore, fish-based food, as fish burgers, represent a valid solution to accomplish consumer preference with products that have high nutritional value and are also very convenient, being ready-to-cook [16,17]. Generally, raw materials, processing technologies, storage conditions, enzymes of fish, and microflora are mainly responsible for fresh fish unacceptability [18]. Even though various preservation strategies are proposed in the literature to preserve seafood products [19–21], the most diffused approach for fresh fish burgers is based on the adoption of natural compounds, properly encapsulated or combined with modified atmosphere conditions [16,17,22–24] or enclosed in edible films [25].

Therefore, in the perspective of a more sustainable food production, the current study, for the first time explored the possibility to adopt all parts of prickly pear fruit for preserving fish burger quality during storage. Specifically, the pulp and peels of prickly pears were dehydrated, ground and included in the fish formulation according to the zero-waste approach, i.e., without producing any waste. The study demonstrated that the newly developed fish burgers, enriched with bioactive compounds from prickly pears, not only exerted good antimicrobial activity against the traditional fish spoilage, but they were also more appreciated than the control sample in terms of sensory quality, thus demonstrating the feasibility of the sustainable approach.

2. Materials and Methods

2.1. Prickly Pears

Red prickly pear fruits (*Opuntia ficus-indica*, (L.) *Miller*, cultivar *Sanguigna*) were kindly provided by a local dealer (Manfredonia, Puglia, Italy), in mid-September 2020. At the laboratory, the prickly pears were washed several times with water to remove any kind of residue and dipped into chlorinated water (20 mL·L^{-1}) for 1 min. Then, the fruits were rinsed with water and air dried. After that, the peel was manually separated from the pulp using a knife and cut into strips, while the pulp, along with the seeds, was cut into small pieces. The peel and pulp were dried at 37 °C for a week using a vacuum dehydrator (Melchioni-Babele, Milan, Italy), and then milled in a lab-grinder to obtain a fine powder (500 micrometers). The two powders (pulp moisture 11.24 ± 0.76% and peel moisture 19.04 ± 0.20%), were stored separately in polyethylene bags at −20 °C until their use. In particular, 19520.1 g of fresh prickly pears was processed; they were composed by 8429.29 g of peel and 11090.81 g of pulp. They were dehydrated and 3049.89 g of dry prickly pear powder was obtained, composed by 1739.72 g pulp, and 1310.17 g of peel. To use all parts of prickly pears, according to the zero-waste approach, for 1 g of prickly pear powder, 0.57 g of pulp and 0.43 g of peel were used.

2.2. Fish Burger Formulation

Frozen cod (*Gadus morhua*) fillets were purchased from a local market (Ipercoop, Foggia, Italy) and were minced with a knife in small pieces. The formulation of the control burger and of the three active samples enriched with the prickly pear powder (named CNT, ACT-2.5, ACT-7.5 and ACT-12.5), is described in Table 1. The fish burgers were obtained by first mixing the flours (i.e., potato starch and potato flakes), salt, and the prickly pear powder, then adding the extra-virgin olive oil, and finally adding the minced cod fillets. Thereafter, the prepared mixture was shaped by means of a metal shaper to obtain fish burgers of 1.5 cm thick, and 5 cm in diameter. Three independent replicates were prepared for each formulation. All investigated burgers were packed into polyethylene bags, sealed and stored at 4 °C.

Table 1. Formulation of the cod fish burgers with and without prickly pear powder addition.

Ingredients	CNT		ACT-2.5		ACT-7.5		ACT-12.5	
	Weight [g]	Weight [%]	Weight [g]	Weight [%]	Weight [g]	Weight [%]	Weight [g]	Weight [%]
Cod fillet	36.98	73.9	36.98	70.4	36.98	64.3	36.98	59.1
Extra-virgin olive oil	4.62	9.2	4.62	8.8	4.62	8.0	4.62	7.4
Potato starch	4.62	9.2	4.62	8.8	4.62	8.0	4.62	7.4
Potato flakes	3.7	7.4	3.7	7.0	3.7	6.4	3.7	5.9
Salt	0.1	0.2	0.1	0.2	0.1	0.2	0.1	0.2
Prickly pear powder	-	-	2.5	4.8	7.5	13.0	12.5	20.0

2.3. Prickly Pear Powder Antibacterial Activity: In Vitro Test

The antimicrobial activity of prickly pear peel and pulp powders was evaluated by an in vitro test. To this aim, two strains of *Pseudomonas* spp. (*P. fluorescens* and *P. putida*) were used as target microorganisms, stored at −20 °C as stock cultures. The exponentially growing cultures were obtained in Plate Count Broth (PCB, tryptone 5 g/L, glucose 1 g/L and yeast extract 2.5 g/L, Oxoid) at 25 °C for 24 h. After that, a cocktail of the two strains was diluted with 0.9% NaCl to obtain approximately 10^3 CFU/mL. The PCB inoculated with the microbial cocktail was placed in several tubes. Every tube also contained 2.5% and 5% of prickly pear peel powder, 2.5%, and 5% of prickly pear pulp powder for the active samples, and no powder was added for the control one. The pulp and the peel allowed to settle in each tube. All tubes were incubated at 25 °C for 72 h. Microbiological analyses were performed after 0, 4, 24, 48, and 72 h taking aliquots of 1 mL from each tube. After appropriate dilutions with 0.9% NaCl, the samples were plated on Pseudomonas Agar Base (PAB, Oxoid), added with cetrimide fucidin cephaloridine (CFC) selective supplement and incubated at 25 °C for 48 h. All analyses were performed twice on two different samples.

2.4. Microbiological Analyses and pH Evaluation

The microbiological analyses were carried out not only on the burgers but also on the powder because peel and pulp dehydration at 37 °C took about a week, and therefore a possible contamination occurred. To the aim, 10 g of powder and 20 g of burger were aseptically transferred in sterile stomacher bags, diluted with 0.9% NaCl solution and homogenized with Stomacher LAB Blender 400 (Pbi International, Milan, Italy). Subsequently, decimal dilutions of homogenate samples were conducted using the same diluent and the dilutions were plated on appropriate media in Petri dishes. In particular, for the fish burgers, the media and the conditions used were: Plate Count Agar (PCA, Oxoid) incubated at 30 °C for 48 h and 5 °C for 10 days for the enumeration of mesophilic and psychrotrophic bacteria, respectively; Pseudomonas Agar Base (PAB, Oxoid), added with cetrimide fucidin cephaloridine (CFC) selective supplement and incubated at 25 °C for

48 h to enumerate *Pseudomonas* spp.; pour plated Iron Agar (IA) incubated at 25 °C for 3 days, for hydrogen sulfide producing bacteria (HSPB); pour plated IA, supplemented with 5 g/L NaCl and incubated at 15 °C for 7 days, for psychrotolerant and heat-labile aerobic bacteria (PHAB); Violet Red Bile Glucose Agar (VRBGA, Oxoid) incubated at 37 °C for 24 h for *Enterobacteriaceae*; de Man Rogosa Sharpe Agar (MRS, Oxoid), supplemented with cycloheximide (0.1 g/L Sigma) incubated at 37 °C for 48 h for Lactic Acid Bacteria (LAB). For the active powder the bacteria and the conditions used for their count were the same used for the burgers, except the PHAB detection because this microbial group was searched for exclusively in the fish products.

The pH of each homogenized sample was measured by a pH meter (Crison, Barbellona, Spain), after appropriate calibration.

2.5. Color Evaluation

Instrumental color readings of the prepared fish burgers were measured with Chromameter CR-400 colorimeter (Minolta Chromameter, Japan) after appropriate calibration, using a reference white tile. The color measurements were described in terms of Lightness (L^*), redness (a^*) and yellowness (b^*) space values. Color measurements were made on the surface of each sample. For all investigated burgers four random readings were performed.

2.6. Sensory Evaluation

The sensory evaluation was performed by seven experienced panelists, researchers of the University of Foggia, selected several years prior to this research for their sensory skills to estimate fish attributes. For the current study, in a 3-h session, panelist reliability was assessed and sensory parameters to be taken into account were defined. Since sensory parameters could be differently perceived on raw and cooked burgers, the sensory evaluation was conducted on both, as also reported in other study [23]. Therefore, at each sampling time (0, 2, 5, 7, 9, 12, 14, 16), the CNT and the three active cod burgers (ACT-2.5; ACT-7.5 and ACT-12.5) were cooked in an electric convention oven (H2810, Hugin, Milan, Italy) at 180 °C for 20 min. For the analysis, each sample, both raw and cooked, was codified with three-digit code and offered to the panelists in individual cabin under controlled conditions of light, temperature and humidity. The panelists were asked to evaluate the color, odor, texture, and then to give an overall quality judgment of the fish burgers. The texture was judged by considering the force exerted for cutting the product with a knife [16]. A 9-point scale was used to quantify each attribute, where a score of 9 corresponded to "very good quality", 7–8 to "good quality" and 6 to "sufficient quality". The value of 5 represented the acceptability threshold, while scores from 4 to 1 corresponded to "unacceptable quality".

2.7. Statistical Analysis

Tests were performed on duplicate batches. All experimental data are the average of three replicates. The results are presented as means ± Standard Deviation (SD) and graphically reported. Statistical significance was determined by one-way analysis of variance (ANOVA). Duncan's multiple range test, with the option of homogeneous groups ($p \leq 0.05$), was performed to determine significant differences among samples. To the aim, STATISTICA 7.1 for Windows (StatSoft, Inc., Tulsa, OK, USA) was used.

3. Results and Discussion

3.1. Effects of Prickly Pear Powder on Pseudomonas spp. by In Vitro Test

A preliminary in vitro test was performed on target bacteria (*Pseudomonas* spp.), before testing the active powder on fish burgers. The results are shown in Table 2. As it can be seen, there is a significant difference between the control sample and the active solutions, already after 24 h from the microbial inoculation. In particular, from the above mentioned data, it can be inferred that the Ctrl had a steady growth during all 72 h of the test, whereas for the active samples a reduction of the *Pseudomonas* spp. viable cell concentration was observed. It is worth noting that there is no difference between the investigated active samples, neither in terms of quantity used (i.e., 2.5% or 5%) nor in terms of the type of by-product (i.e., peel or pulp powder). As a matter of fact, all the concentrations chosen for the in vitro test showed a substantial antimicrobial activity on the target bacteria already after one day.

Table 2. Evolution of the *Pseudomonas* spp. viable cell concentration after inoculation of control and prickly pear peel (2.5% and 5%) and pulp (2.5% and 5%) powder. Data indicate means ± SD.

	log CFU/mL				
Sample Time (h)	0	4	24	48	72
Ctrl	2.96 ± 0.11	3.46 ± 0.09	8.82 ± 0.39	9.41 ± 0.17	9.49 ± 0.23
Peel 2.5%	3.12 ± 0.01	3.62 ± 0.09	1.30 ± 0.43	<10	<10
Peel 5%	3.26 ± 0.08	4.05 ± 0.14	2.09 ± 0.12	<10	<10
Pulp 2.5%	3.57 ± 0.21	3.80 ± 0.14	1.35 ± 0.49	<10	<10
Pulp 5%	3.51 ± 0.01	3.54 ± 0.09	2.76 ± 0.09	<10	<10

3.2. Microbial Contamination of Prickly Pear Powder

A preliminary microbiological analysis on the powder was carried out to assess whether prolonged dehydration at low temperature could have affected the microbiological stability. The results of this test are reported in Table 3. It can be noticed that for both the peel and pulp powders, only psychrotrophic bacteria were absent. The total mesophilic count (MES) and the lactic acid bacteria (LAB) recorded higher values compared to the other investigated microorganisms. Therefore, these results suggest that prickly pear dehydration at 37 °C promoted spoilage growth, most probably because the process took about one week. Due to the recorded powder contamination, antimicrobial effects of powder on fish burgers were assessed by comparing growth rate of main spoilage microorganisms, instead of using the values of specific viable cell concentrations.

Table 3. Contamination of prickly pear peel and pulp powder (log CFU/g).

Sample	PSY	MES	PSE	ENTER	LAB	SHPB
Peel	-	7.14 ± 0.20	3.43 ± 0.10	4.71 ± 0.11	7.07 ± 0.10	6.02 ± 0.53
Pulp	-	7.03 ± 0.04	3.56 ± 0.04	3.35 ± 0.16	7.12 ± 0.16	5.15 ± 0.27

Data indicate means ± SD. PSY = Total Psycrhrotrophic bacteria < 10^2 CFU/g; MES = Total Mesophilic bacteria; PSE = *Pseudomonas* spp.; ENTER = Enterobacteriaceae; LAB = Lactic Acid Bacteria; SHPB = Hydrogen Sulfide producing bacteria.

3.3. Effects of Prickly Pear Powder on Microbiological Quality of Cod Fish Burgers

As reported in Section 2, the effects of prickly pear powder (i.e., 2.5 g, 7.5 g, and 12.5 g) on the microbial quality decay of fish burgers during refrigerated storage were assessed. Spoiled fish products are characterized by development of fishy, rotten H_2S off-odors and unpleasant flavors, due to growth of specific spoilage microorganisms [25]. It is important to highlight that seafood deterioration is mainly caused by high microbial growth, that consequently provokes unpleasant chemical compounds production [26]. For this reason, *Pseudomonas* spp., hydrogen sulfide producing bacteria (HSPB), psychrotolerant and heat-labile aerobic bacteria (PHAB) were monitored as main spoilage groups.

Figure 1 shows the evolution of *Pseudomonas* spp. viable cell concentration of the investigated fish burgers (i.e., CNT, ACT-2.5, ACT-7.5, and ACT-12.5). As it may be seen, during 16 days of storage, a decrease in the viable cell concentration (negative growth rate) was observed for the two samples ACT-7.5 and ACT-12.5. Despite the initial contamination, the addition of 7.5 g and 12.5 g of prickly pear powder had a significant effect on the growth of *Pseudomonas* spp., confirming data from the in vitro test (see Table 2). As one would expect, the CNT sample showed a rapid increase of microbial load during the first week of storage, which remained quite stable up to the end of the observation period. As shown in Figure 1, the ACT-2.5 sample had a trend similar to CNT, thus suggesting that at the lowest concentration, prickly pear powder had a minimal antimicrobial effect on *Pseudomonas* spp. proliferation. These results are in accordance with those of Ennouri et al. [27], who proved the antibacterial activity of the extract from prickly pear against *P. aeruginosa*. Furthermore, Palmeri et al. [12] also found that the water-based extract of prickly pear was effective in reducing the count of *Pseudomonas* spp. on sliced beef.

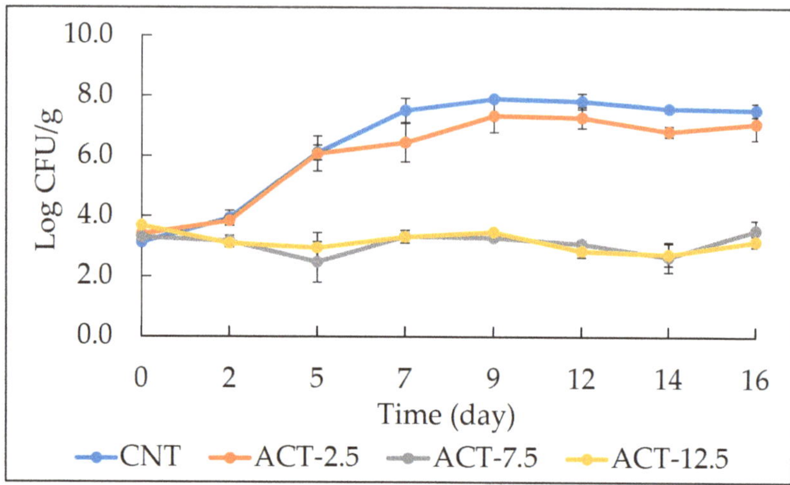

Figure 1. The evolution of *Pseudomonas* spp. viable cell concentration in fish burgers during 16 days of storage at 4 °C. Data indicate means ± SD. CNT: fish burger without prickly pear powder; ACT-2.5: fish burger enriched with 2.5 g of prickly pear powder; ACT-7.5: fish burger enriched with 7.5 g of prickly pear powder; ACT-12.5: fish burgers enriched with 12.5 g of prickly pear powder.

Figure 2 exhibits the evolution of total psychrotrophic bacterial load during refrigerated storage. As can be seen, ACT-12.5 had shown the best antimicrobial effect throughout the storage period. It is worth noting that, despite powder contamination (see Table 3), a significant antimicrobial efficacy was found. In fact, a slight decrease in the viable cell concentration, followed by a gradual increase, was observed for the ACT-12.5 fish burger. The CNT, the ACT-2.5 and the ACT-7.5 samples have had a steady growth throughout the entire observation period. Among these last three samples, ACT-7.5 showed the best antimicrobial effect (i.e., the lowest growth rate), whereas ACT-2.5 and CNT showed a similar trend. These results suggest that prickly pear powder is not able to slow down the growth of psychrotrophic bacteria if used at low concentrations.

To make a comparison with literature, it must be observed that the results obtained in our experimental plan for the CNT sample are similar to those found in other studies dealing with fresh fish burgers [23,28], whereas, as regard active samples, Panza et al. [29] can be cited because these authors obtained a similar trend of psychrotrophic bacteria using pomegranate by-products to extend the shelf life of breaded cod sticks.

The PHAB viable cell concentration plotted as a function of storage time is shown in Figure 3 for all fish burgers investigated in this study. The highest powder amount (12.5 g of active powder) was proven to be the most effective against this spoilage group. As a matter of fact, ACT-12.5 showed any microbial growth throughout the refrigerated storage. In the case of ACT-7.5 sample the microbial detection highlighted the same effects recorded in the previous sample (ACT-12.5), but there was a slight increase only on the last day of storage. The figure also shows that the CNT and the ACT-2.5 presented a very comparable trend, with a visible and rapid microbial growth from the second day of storage up to the end of the observation period.

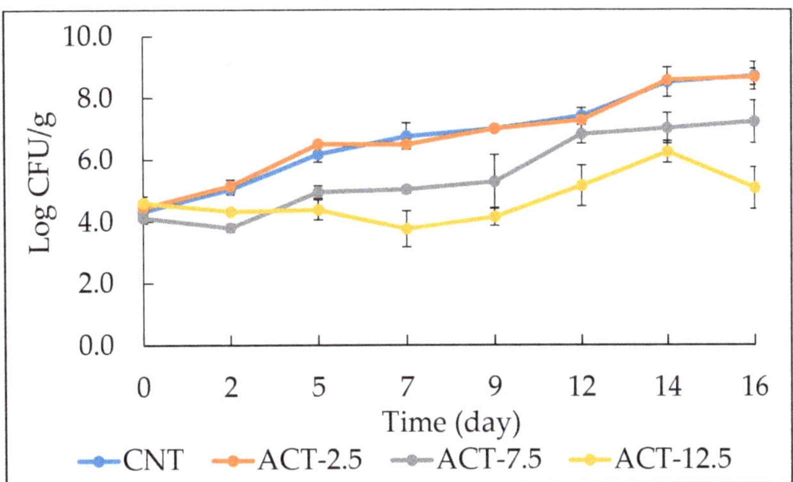

Figure 2. The evolution of total psycrhrotrophic bacteria in fish burgers during 16 days of storage at 4 °C. Data indicate means ± SD. CNT: fish burger without prickly pear powder; ACT-2.5: fish burger enriched with 2.5 g of prickly pear powder; ACT-7.5: fish burger enriched with 7.5 g of prickly pear powder; ACT-12.5: fish burgers enriched with 12.5 g of prickly pear powder.

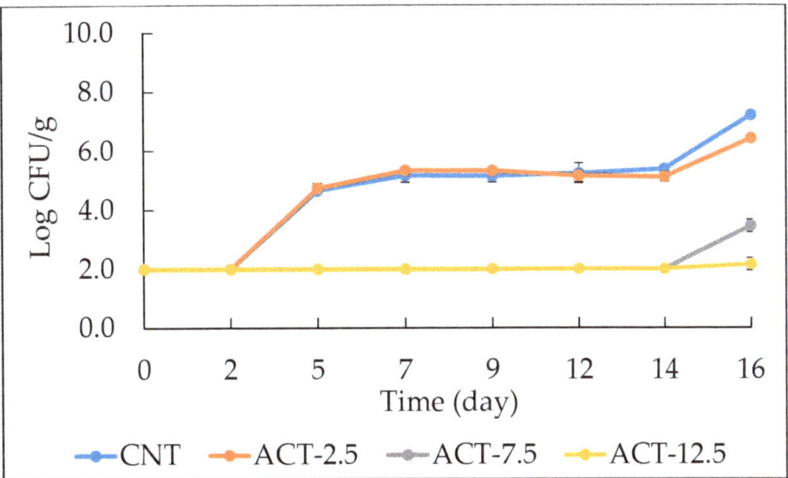

Figure 3. The evolution of psychrotolerant and heat-labile aerobic (PHAB) viable cell concentration in fish burgers during 16 days of storage at 4 °C. Data indicate means ± SD. CNT: fish burger without prickly pear powder; ACT-2.5: fish burger enriched with 2.5 g of prickly pear powder; ACT-7.5: fish burger enriched with 7.5 g of prickly pear powder; ACT-12.5: fish burgers enriched with 12.5 g of prickly pear powder.

Figure 4 shows the microbial load evolution during the storage of the following groups: total mesophilic bacteria (Figure 4a), *Enterobacteriaceae* (Figure 4b), LAB (Figure 4c) and HSPB (Figure 4d). Despite the aforementioned powder contamination (see Table 3), for all the bacterial groups the antimicrobial effect of the prickly pear powder is evident when the growth rate of each sample is taken into account. As one would expect, the ACT-12.5 sample had the best antimicrobial effect considering that its growth rate was negligible and in some cases it was negative. Although the antimicrobial activity of the ACT-7.5

sample was evident, it was found to be slightly lower than that observed for the ACT-12.5 fish burger. For both CNT and ACT-2.5 a rapid increase in viable cell concentration was observed, visible for each microbial group (a–d). Regarding the growth rate, the ACT-2.5 sample had a lower value if compared to CNT, thus suggesting that also at the lowest concentration, prickly pear powder exerted a slight antimicrobial effect on these microbial groups. Similar trends were found by Nisar et al. [30], who used rosemary oil as natural preservative for bream fillets.

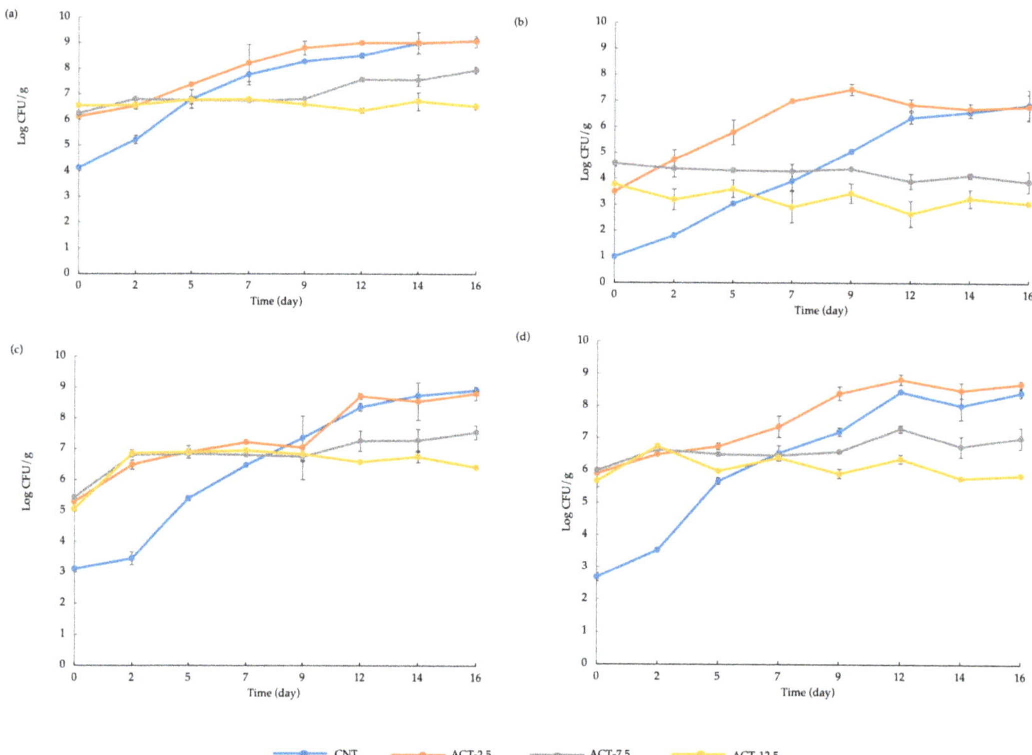

Figure 4. The evolution of total mesophilic bacteria (**a**) *Enterobacteriaceae* (**b**) Lactic acid bacteria (**c**) and Hydrogen Sulfide producing bacteria (**d**) in fish burgers during 16 days of storage at 4° C. Data indicate the mean ± SD. CNT: fish burger without prickly pear powder; ACT-2.5: fish burger enriched with 2.5 g of prickly pear powder; ACT-7.5: fish burger enriched with 7.5 g of prickly pear powder; ACT-12.5: fish burgers enriched with 12.5 g of prickly pear powder.

3.4. Effects of Prickly Pear Powder on pH of Cod Fish Burgers

The evolution during refrigerated storage of fish burger pH was reported in Figure 5. Initially, pH was similar for all the examined samples (around 7), in accordance with values also reported by other authors [28,31,32]. During time, the pH of CNT, ACT-7.5, and ACT-12.5 samples was almost constant throughout the storage period (6.53, 6.24 and 6.31 respectively), although the values of the two active burgers were slightly lower than those of the CNT. As long as the ACT-2.5 sample is concerned, pH values slightly decreased during time, most probably due to the high microbial proliferation occurred in this samples, in particular for the high lactic acid bacteria proliferation [33,34].

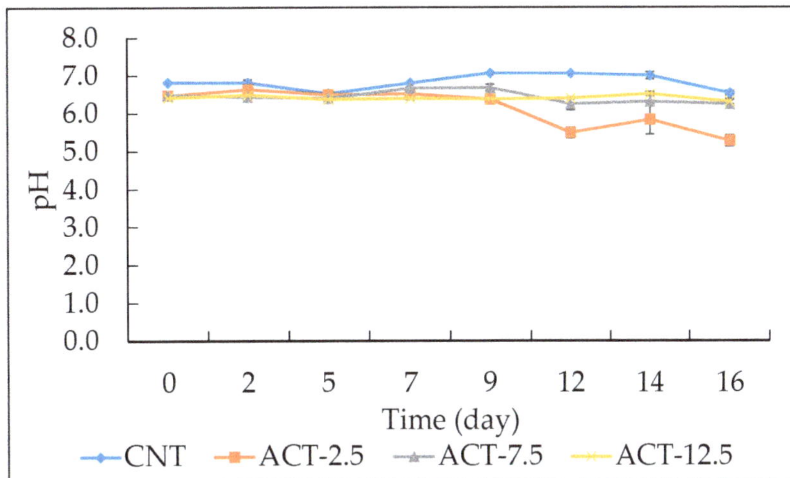

Figure 5. Trend of pH values of fish burgers during 16 days of storage at 4 °C. Data indicate means ± SD. CNT: fish burger without prickly pear powder; ACT-2.5: fish burger enriched with 2.5 g of prickly pear powder; ACT-7.5: fish burger enriched with 7.5 g of prickly pear powder; ACT-12.5: fish burgers enriched with 12.5 g of prickly pear powder.

3.5. Effects of Prickly Pear Powder on Sensory Quality of Cod Fish Burgers

The scores for color, odor, and texture at the first and the last day of storage are presented in Table 4, for both raw and cooked fish burgers. Adding prickly pear powder, an aromatic and pleasant odor was perceived by the panelists, associated to the volatile compounds of prickly pears [7–11]. This perception was observed from the beginning and up to the last day of storage, for both raw and cooked fish products. These results are proven by the scores obtained from the panel, which were above the acceptability threshold (score = 5). Moreover, for ACT-7.5 and ACT-12.5, the presence of the powder led to the detection of any kind of fish odor, even at the last day of storage. An opposite behavior was found for the CNT samples that showed a high level of off-odors, especially on the 16th day of storage. Regarding the texture, in all the investigated samples, the values decreased during the storage period, thus proving that hardness and structure of the fish were lost during time. However, a statistically significant effect ($p > 0.05$) of prickly pear powder was observed. In particular, as the powder increased, the detrimental effect of storage time on the sample texture was less evident. This implies that the use of the powder could slow down texture decay during time. As long as the color attribute is concerned, a trend similar to that of the other two sensory attributes was observed. In particular, on the last day of storage, CNT sample recorded the worst score, ACT-2.5 was slightly better than the CNT, ACT-7.5 and ACT-12.5 obtained the best scores, and the differences between them were not statistically significant ($p < 0.05$).

The color change of the investigated fish burgers was also assessed by the colorimeter. The results are listed in Table 5, where the values are expressed as lightness (L*), redness (a*), and yellowness (b*). Data in Table 5 highlight that addition of prickly pear powder reduced the lightness of the burgers if compared to the CNT samples. Looking at the first column of Table 5, it can be also observed that lightness decreased as the powder concentration increased. This finding is not surprising, considering the dark color of the powder. In terms of the effects of time on color parameters, no great influence can be underlined. From data it's possible to infer that lightness only in some cases slightly rose over storage, specifically for ACT-2.5 and ACT-7.5 [28].

Table 4. Scores of sensory attributes (color, odor and texture) of raw and cooked fish burgers at 0 and 16 days of storage.

Samples		Storage Time (day)			
		Raw Fish Burger		Cooked Fish Burger	
		0	16	0	16
Color	CNT	8.1 ± 0.25 ^{b}	2.9 ± 0.25 ^{c}	8.3 ± 0.29 ^{b}	4.1 ± 0.25 ^{b}
	ACT-2.5	9.0 ± 0.10 ^{a}	4.5 ± 0.58 ^{b}	9.0 ± 0.10 ^{a}	4.8 ± 0.29 ^{b}
	ACT-7.5	9.0 ± 0.10 ^{a}	6.1 ± 0.48 ^{a}	9.0 ± 0.10 ^{a}	6.3 ± 0.87 ^{a}
	ACT-12.5	9.0 ± 0.10 ^{a}	6.4 ± 0.75 ^{a}	9.0 ± 0.10 ^{a}	7.0 ± 0.41 ^{a}
Odor	CNT	9.0 ± 0.10 ^{a}	2.4 ± 0.25 ^{d}	9.0 ± 0.10 ^{a}	3.5 ± 0.58 ^{b}
	ACT-2.5	9.0 ± 0.10 ^{a}	3.6 ± 0.48 ^{c}	9.0 ± 0.10 ^{a}	3.6 ± 0.25 ^{b}
	ACT-7.5	9.0 ± 0.10 ^{a}	6.0 ± 0.29 ^{b}	9.0 ± 0.10 ^{a}	6.4 ± 0.48 ^{a}
	ACT-12.5	9.0 ± 0.10 ^{a}	6.9 ± 0.25 ^{a}	9.0 ± 0.10 ^{a}	7.0 ± 0.41 ^{a}
Texture	CNT	8.1 ± 0.25 ^{b}	3.6 ± 0.25 ^{c}	8.3 ± 0.29 ^{a}	4.0 ± 0.41 ^{c}
	ACT-2.5	8.9 ± 0.25 ^{a}	5.8 ± 0.50 ^{b}	8.8 ± 0.29 ^{a}	4.8 ± 0.29 ^{b}
	ACT-7.5	9.0 ± 0.10 ^{a}	6.5 ± 0.41 ^{a,b}	8.9 ± 0.25 ^{a}	7.0 ± 0.41 ^{a}
	ACT-12.5	8.9 ± 0.25 ^{a}	7.0 ± 0.71 ^{a}	8.6 ± 0.48 ^{a}	7.3 ± 0.29 ^{a}

$^{a–d}$ Data indicate means ± SD. For each sensory attribute, the values marked with different superscript letters in the column are significantly different ($p < 0.05$). CNT: fish burger without prickly pear powder; ACT-2.5: fish burger enriched with 2.5 g of prickly pear powder; ACT-7.5: fish burger enriched with 7.5 g of prickly pear powder; ACT-12.5: fish burgers enriched with 12.5 g of prickly pear powder.

Table 5. The effect of prickly pear enrichment on the color parameters of fish burgers.

Samples	L*		a*		b*	
	0	16	0	16	0	16
CNT	67.73 ± 1.87 ^{a,A}	68.68 ± 1.97 ^{a,A}	−2.79 ± 0.81 ^{b,A}	−2.42 ± 0.78 ^{d,A}	16.48 ± 1.43 ^{b,B}	18.12 ± 1.64 ^{c,A}
ACT-2.5	52.33 ± 1.56 ^{b,B}	57.07 ± 1.64 ^{b,A}	11.70 ± 1.78 ^{a,A}	5.69 ± 0.59 ^{c,B}	22.07 ± 1.80 ^{a,B}	27.63 ± 1.62 ^{a,A}
ACT-7.5	43.02 ± 1.04 ^{c,B}	46.76 ± 1.70 ^{c,A}	14.14 ± 1.95 ^{a,A}	9.15 ± 0.68 ^{a,B}	17.32 ± 1.41 ^{b,B}	22.41 ± 1.11 ^{b,A}
ACT-12.5	41.73 ± 0.39 ^{c,A}	42.50 ± 1.66 ^{d,A}	14.45 ± 0.73 ^{a,A}	7.89 ± 0.58 ^{b,B}	15.29 ± 0.81 ^{b,A}	16.42 ± 1.32 ^{c,A}

Data indicate means ± SD. Values marked with different superscript letters (a–d) in the column and superscript uppercase letters (A, B) in the row are significantly different ($p < 0.05$). CNT: fish burger without prickly pear powder; ACT-2.5: fish burger enriched with 2.5 g of prickly pear powder; ACT-7.5: fish burger enriched with 7.5 g of prickly pear powder; ACT-12.5: fish burgers enriched with 12.5 g of prickly pear powder. L* = lightness (0 = darkness, 100 = lightness); a* = redness (+60 = red, −60 = green); b* = yellowness (+60 = yellow, −60 = blue).

The redness parameter (a*) was completely absent in CNT burgers during the entire storage period, while for the active samples, there was a significant reduction between the first and the last day of storage, more marked for the ACT-12.5. This color modification is caused by the migration of the prickly pear pigments from the powder to the minced cod fillets [35].

As far the yellowness parameter is concerned (b*), there were slight differences between CNT and active samples at both the beginning and at the end of the observation period. An increase in the b* value was observed for all samples between zero and 16 days of storage, thus demonstrating that fish decay is strictly linked to changes in yellowness parameters, regardless the by-product concentration. As a result, the use of the powder significantly affected the visual quality of fish burgers, as confirmed from data recorded on the color parameter during the panel test.

The evolution of fish burger overall quality during 16 days of storage is shown in Figure 6a,b. A more pronounced reduction of the fish burger sensory quality was observed for the investigated raw samples, being the decay of the raw products faster than that of the cooked ones. As can be inferred from data shown in both graphs of Figure 6, CNT and ACT-2.5 had a similar trend, and became unacceptable during time. On the other hand, ACT-7.5 and ACT-12.5 samples showed good overall quality throughout the entire observation period and after two weeks these samples were found still completely acceptable. This may be due to the antimicrobial effect exerted by prickly pear powder, and in particular to the aromatic compounds that preserved the odor. Therefore, it is worth noting that the prickly pear powder added to the fish burgers did not worsen sensory quality, rather it improved the fish characteristics, especially in the raw samples (Figure 6a). In fact, already at the beginning of the sensory evaluation (time 0), the panelists appreciated the active samples more than the control. To sum up, the findings previously recorded in Table 4, dealing with specific sensory attributes, are in accordance with data on the general acceptance of the products shown in Figure 6.

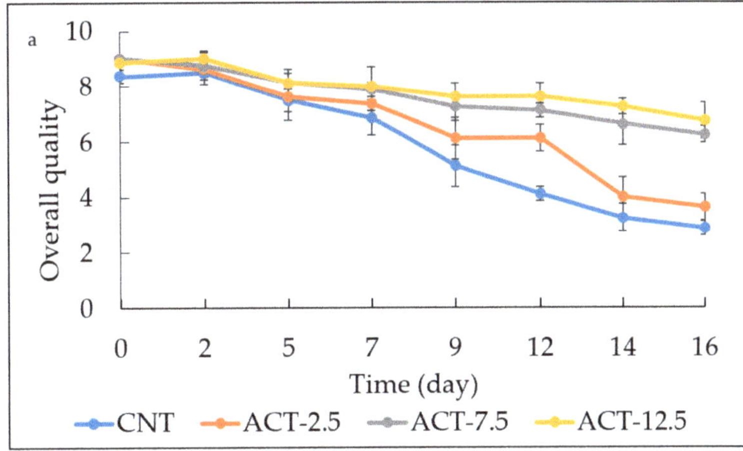

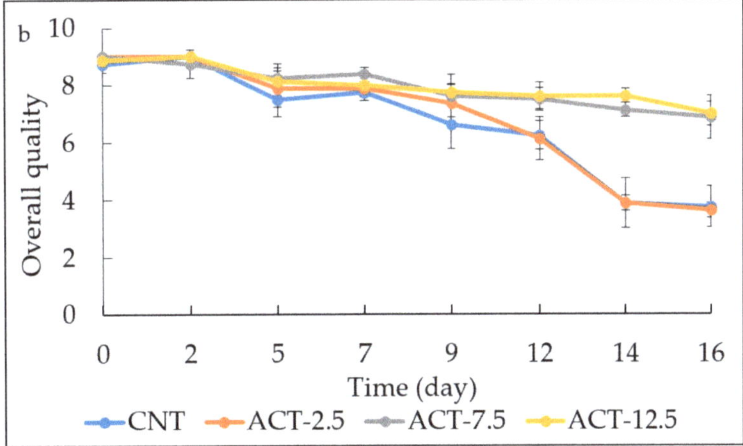

Figure 6. Evolution of Overall quality of both raw (**a**) and cooked (**b**) fish burgers during 16 days of storage. Data indicate means ± SD. CNT: fish burger without prickly pear powder; ACT-2.5: fish burger enriched with 2.5 g of prickly pear powder; ACT-7.5: fish burger enriched with 7.5 g of prickly pear powder; ACT-12.5: fish burgers enriched with 12.5 g of prickly pear powder.

4. Conclusions

The current study aimed to improve the storability of refrigerated cod fish burgers in accordance with the zero-waste concept. In fact, all the prickly pear parts were dehydrated, reduced in powder, and then used as food additives in the fish formulation, in a proportion that respected the zero-waste approach. An in vitro test was first run to assess the antimicrobial efficacy of this fruit powder. The results showed that both the concentrations chosen (2.5 and 5%) in this preliminary test had a marked antimicrobial activity on target bacteria, already after 24 h. The prickly pear powder was therefore added at three different amounts (2.5 g, 7.5 g and 12.5 g) to cod fish burgers, that were stored at 4 °C to assess powder efficacy in slowing down microbial growth. The results showed that the lowest prickly pear powder concentration used in this study did not affect the microbial growth at a great extent. Conversely, the other two tested concentrations, and especially the highest one, had a remarkable effect on the growth kinetic of spoilage microorganisms. In fact, in most cases a reduction of the growth rate was observed, if compared to the control sample. In few cases the growth rate was close to zero (i.e., total mesophilic bacteria and HSPB), and in one case (i.e., *Enterobacteriaceae*) a decrease in the viable cell concentration (negative growth rate) during refrigerated storage was observed. As long as the sensory quality decay is concerned, the results obtained in this study indicated that the prickly pear powder at two highest concentrations significantly slowed down the sensory quality decay during time of both raw and cooked cod fish burgers. This is most probably due to the antimicrobial activity of the powder and its aromatic compounds that improved the perceived odor. As a fact, the two burgers with 7.5 g and 12.5 g powder addition remained completely acceptable during the entire observation period. To sum up, due to the recorded results, the recycle of prickly pears seems interesting. The research needs to further explore the topic in terms of energy costs, emissions, etc., because drying of prickly pears is an energy-intensive operation. The use of renewable energy could be a starting point because companies are increasingly moving towards the use of green energy. Certainly, more in-depth studies would be needed for recommending zero-waste as a beneficial approach.

Author Contributions: Conceptualization, M.A.D.N. and A.C.; methodology, M.A.D.N., A.C.; formal analysis, F.D. and V.L.; data curation, M.A.D.N. and A.C.; writing—original draft preparation, F.D.; writing—review and editing, M.A.D.N. and A.C.; supervision A.C. All authors have read and agreed to the published version of the manuscript.

Funding: This research received no external funding.

Data Availability Statement: The raw data will be made available upon request.

Conflicts of Interest: The authors declare no conflict of interest.

References

1. Aka, S.; Buyukdag, N. How to prevent food waste behaviour? A deep empirical research. *J. Retail. Consum. Serv.* **2021**, *61*, 102560. [CrossRef]
2. Song, Q.; Li, J.; Zeng, X. Minimizing the increasing solid waste through zero waste strategy. *J. Clean. Prod.* **2015**, *104*, 199–210. [CrossRef]
3. Kumar, K.; Srivastav, S.; Sharanagat, V.S. Ultrasound assisted extraction (UAE) of bioactive compounds from fruit and vegetable processing by-products: A review. *Ultrason. Sonochem.* **2021**, *70*, 105325. [CrossRef]
4. Dilucia, F.; Lacivita, V.; Conte, A.; Del Nobile, M.A. Sustainable Use of Fruit and Vegetable By-Products to Enhance Food Packaging Performance. *Foods* **2020**, *9*, 857. [CrossRef]
5. Karimi, A.; Kazemi, M.; Amiri Samani, S.; Simal-Gandara, J. Bioactive compounds from by-products of eggplant: Functional properties, potential applications and advances in valorization methods. *Trends Food Sci. Technol.* **2021**, *112*, 518–531. [CrossRef]
6. Andreu-Coll, L.; Cano-Lamadrid, M.; Sendra, E.; Carbonell-Barrachina, Á.; Legua, P.; Hernández, F. Fatty acid profile of fruits (pulp and peel) and cladodes (young and old) of prickly pear [*Opuntia ficus-indica* (L.) Mill.] from six Spanish cultivars. *J. Food Compos. Anal.* **2019**, *84*, 103294. [CrossRef]
7. Bouazizi, S.; Montevecchi, G.; Antonelli, A.; Hamdi, M. Effects of prickly pear (*Opuntia ficus-indica* L.) peel flour as an innovative ingredient in biscuits formulation. *LWT Food Sci. Technol.* **2020**, *124*, 109155. [CrossRef]
8. Cardador-Martínez, A.; Jimenez Martinez, C.; Sandoval, G. Revalorization of cactus pear (*Opuntia* spp.) wastes as a source of antioxidants. *Food Sci. Technol.* **2011**, *31*, 782–788. [CrossRef]
9. Palmeri, R.; Parafati, L.; Arena, E.; Grassenio, E.; Restuccia, C.; Fallico, B. Antioxidant and antimicrobial properties of semi-processed frozen prickly pear juice as affected by cultivar and harvest time. *Foods* **2020**, *9*, 235. [CrossRef] [PubMed]
10. Karabagias, V.K.; Karabagias, I.K.; Prodromiti, M.; Gatzias, I.; Badeka, A. Bio-functional alcoholic beverage preparation using prickly pear juice and its pulp in combination with sugar and blossom honey. *Food Biosci.* **2020**, *35*, 100591. [CrossRef]

11. Valero-Galván, J.; González-Fernández, R.; Sigala-Hernández, A.; Núnez-Gastélum, J.A.; Ruiz-May, E.; Rodrigo-García, J.; Larqué-Saavedra, A.; del Rocío Martínez-Ruiz, N. Sensory attributes, physicochemical and antioxidant characteristics, and protein profile of wild prickly pear fruits (*O. macrocentra* Engelm., *O. phaeacantha* Engelm., and *O. engelmannii* Salm-Dyck ex Engelmann.) and commercial prickly pear fruits (*O. ficus-indica* (L.) Mill.). *Food Res. Int.* **2021**, *140*, 109909. [CrossRef]
12. Palmeri, R.; Parafati, L.; Restuccia, C.; Fallico, B. Application of prickly pear fruit extract to improve domestic shelf life, quality and microbial safety of sliced beef. *Food Chem. Toxicol.* **2018**, *118*, 355–360. [CrossRef]
13. Melgar, B.; Dias, M.I.; Ciric, A.; Sokovic, M.; Garcia-Castello, E.M.; Rodriguez-Lopez, A.D.; Barros, L.; Ferreira, I. By-product recovery of *Opuntia* spp. peels: Betalainic and phenolic profiles and bioactive properties. *Ind. Crop. Prod.* **2017**, *107*, 353–359. [CrossRef]
14. Chougui, N.; Djerroud, N.; Naraoui, F.; Hadjal, S.; Aliane, K.; Zeroual, B.; Larbat, R. Physicochemical properties and storage stability of margarine containing Opuntia ficus-indica peel extract as antioxidant. *Food Chem.* **2015**, *173*, 382–390. [CrossRef] [PubMed]
15. Parafati, L.; Restuccia, C.; Palmeri, R.; Fallico, B.; Arena, E. Characterization of prickly pear peel flour as a bioactive and functional ingredient in bread preparation. *Foods* **2020**, *9*, 1189. [CrossRef] [PubMed]
16. Danza, A.; Conte, A.; Del Nobile, M.A. Technological options to control quality of fish burgers. *J. Food Sci. Technol.* **2017**, *54*, 1802–1808. [CrossRef] [PubMed]
17. Spinelli, S.; Conte, A.; Lecce, L.; Incoronato, A.L.; Del Nobile, M.A. Microencapsulated propolis to enhance the antioxidant properties of fresh fish burgers. *J. Food Proc. Eng.* **2014**, *38*, 527–535. [CrossRef]
18. Parlapani, F.F. Microbial diversity of seafood. *Curr. Opin. Food Sci.* **2021**, *37*, 45–51. [CrossRef]
19. Houicher, A.; Bensid, A.; Regenstein, J.M.; Ozogul, F. Control of biogenic amine production and bacterial growth in fish and seafood products using phytochemicals as biopreservatives: A review. *Food Biosci.* **2021**, *39*, 100807. [CrossRef]
20. Kontominas, M.G.; Badeka, A.V.; Kosma, I.S.; Nathanailides, C.I. Innovative seafood preservation technologies: Recent developments. *Animals* **2021**, *11*, 92. [CrossRef]
21. Olatunde, O.O.; Benjakul, S. Natural preservatives for extending the shelf-life of seafood: A revisit. *Compr. Rev. Food Sci. Food Saf.* **2018**, *17*, 1595–1612. [CrossRef]
22. Hasani, S.; Ojagh, S.M.; Ghorbani, M.; Hasani, M. Nano-encapsulation of lemon essential oil approach to reducing the oxidation process in fish burger during refrigerated storage. *J. Food Biosci. Technol.* **2020**, *10*, 35–46.
23. Cedola, A.; Cardinali, A.; Del Nobile, M.A.; Conte, A. Fish burger enriched by olive oil industrial by-product. *Food Sci. Nutr.* **2017**, *5*, 837–844. [CrossRef]
24. Del Nobile, M.A.; Corbo, M.R.; Speranza, B.; Sinigaglia, M.; Conte, A.; Caroprese, M. Combined effect of MAP and active compounds on fresh blue fish burger. *Int. J. Food Microbiol.* **2009**, *135*, 281–287. [CrossRef]
25. Albertos, I.; Marrtin-Diana, A.B.; Burón, M.; Rico, D. Development of functional bio-based seaweed (*Himanthalia elongata* and *Palmaria palmata*) edible films for extending the shelf life of fresh fish burgers. *Food Packag. Shelf Life* **2019**, *22*, 100382. [CrossRef]
26. Gram, L.; Dalgaard, P. Fish spoilage bacteria—Problems and solutions. *Curr. Opin. Biotechnol.* **2002**, *13*, 262–266. [CrossRef]
27. Ennouri, M.; Ammar, I.; Khemakhem, B.; Attia, H. Chemical composition and antibacterial activity of *Opuntia Ficus-Indica F. Inermis* (Cactus Pear) Flowers. *J. Med. Food* **2014**, *17*, 908–914. [CrossRef]
28. Rico, D.; Albertos, I.; Martinez-Alvarez, O.; Lopez-Caballero, M.E.; Martin-Diana, A.B. Use of sea fennel as a natural ingredient of edible films for extending the shelf life of fresh fish burgers. *Molecules* **2020**, *25*, 5260. [CrossRef]
29. Panza, O.; Conte, A.; Del Nobile, M.A. Pomegranate by-products as natural preservative to prolong the shelf life of breaded cod stick. *Molecules* **2021**, *26*, 2385. [CrossRef]
30. Nisar, T.; Yang, X.; Alim, A.; Iqbal, M.; Wang, Z.; Guo, Y. Physicochemical responses and microbiological changes of bream (*Megalobrama ambycephala*) to pectin-based coatings enriched with clove essential oil during refrigeration. *Int. J. Biol. Macromol.* **2019**, *124*, 1156–1166. [CrossRef] [PubMed]
31. Mexis, S.F.; Chouliara, E.; Kontominas, M.G. Combined effect of an oxygen absorber and oregano essential oil on shelf life extension of rainbow trout fillets stored at 4 °C. *Food Microbiol.* **2009**, *26*, 598–605. [CrossRef] [PubMed]
32. Gao, M.; Feng, L.; Jiang, T.; Zhu, J.; Fu, L.; Yuan, D.; Li, J. The use of rosemary extract in combination with nisin to extend the shelf life of pompano (*Trachinotus ovatus*) fillet during chilled storage. *Food Control.* **2014**, *37*, 1–8. [CrossRef]
33. Kitundu, E.; Young, O.; Seale, B.; Owens, A. Lactic fermentation of cooked, comminuted mussel, *Perna canaliculus*. *Food Microbiol.* **2021**, *99*, 103829. [CrossRef] [PubMed]
34. Danza, A.; Lucera, A.; Lavermicocca, P.; Lonigro, S.L.; Bavaro, A.R.; Mentana, A.; Centonze, D.; Conte, A.; Del Nobile, M.A. Tuna Burgers Preserved by the Selected *Lactobacillus paracasei* IMPC 4.1 Strain. *Food Bioprocess. Technol.* **2018**, *11*, 1651–1661. [CrossRef]
35. Kharrat, N.; Salem, H.; Mrabet, A.; Aloui, F.; Triki, S.; Fendri, A.; Gargouri, Y. Synergistic effect of polysaccharides, betalain pigment and phenolic compounds of red prickly pear (*Opuntia stricta*) in the stabilization of salami. *Int. J. Biol. Macromol.* **2018**, *111*, 561–568. [CrossRef]

Article

Pomegranate Peel Powder as a Food Preservative in Fruit Salad: A Sustainable Approach

Valentina Lacivita, Anna Lucia Incoronato, Amalia Conte * and Matteo Alessandro Del Nobile

Department of Agricultural Sciences, Food and Environment, University of Foggia, Via Napoli, 25-71121 Foggia, Italy; valentina.lacivita@unifg.it (V.L.); annalucia.incoronato@unifg.it (A.L.I.); matteo.delnobile@unifg.it (M.A.D.N.)
* Correspondence: amalia.conte@unifg.it

Abstract: This study aimed to assess the potential of pomegranate peel powder as a natural preservative. Its effects were tested on fruit salad quality decay during refrigerated storage. Nectarine and pineapple, equally portioned in polypropylene containers and covered with fructose syrup, were closed using a screw cap in air, with and without the addition of a by-product peel powder. Specifically, amounts of 2.5% and 5% (w/v) of pomegranate peel powder were put into each container. Both the microbiological and sensory qualities of the fruit salad were monitored during storage at 5 °C for 28 days. The results demonstrated that the fruit salad with the by-products showed lower counts of total mesophilic bacteria, total psychrotrophic microorganisms, yeasts, and lactic acid bacteria compared to the control, thus confirming the recognized antimicrobial properties of pomegranate peel. The other interesting finding of this study is that the addition of the investigated by-product in fruit salad did not worsen the main sensory attributes of fresh-cut fruit. Therefore, these preliminary results suggest that pomegranate peel powder has potential applications as a natural preservative in the fresh-cut food sector.

Keywords: fresh-cut fruit; pomegranate peel powder; natural preservative; by-product; sustainable approach

Citation: Lacivita, V.; Incoronato, A.L.; Conte, A.; Del Nobile, M.A. Pomegranate Peel Powder as a Food Preservative in Fruit Salad: A Sustainable Approach. *Foods* **2021**, *10*, 1359. https://doi.org/10.3390/foods10061359

Academic Editors: Valeria Rizzo and Muratore Giuseppe

Received: 27 May 2021
Accepted: 8 June 2021
Published: 11 June 2021

1. Introduction

Fruit intake, which is associated with a correct and healthy diet, is increasingly widespread among consumers. Fruit is rich in carbohydrates, minerals, amino acids, vitamins, and other nutrients that bring various benefits to human health [1]. In the last few years, fresh-cut fruit has been popular because it meets consumers' need for fresh, natural, and convenient food. Various formats are available in the refrigerated section, ranging from single fruit to fruit salads. However, these fresh-cut products are all highly perishable [2]. Minimal processing operations, which include peeling, slicing, dicing, etc., can cause damage to fruits' surfaces, thus limiting their shelf life compared to unprocessed whole fruits [3–5]. Damage caused by the cutting process typically occurs in the form of tissue softening, water loss, color change (surface browning), microbial proliferation, and the appearance of unpleasant odors [4–7]. In general, the most common preservation systems for fresh-cut fruit include cold storage, the use of modified atmosphere packaging, coating application, or the addition of synthetic preservatives [8–10]. However, consumers are increasingly inclined to purchase fresh-cut fruit without synthetic additives, as they have a greater awareness of health and food safety. An alternative to synthetic additives could be the use of preservatives of natural origin, such as essential oils, enzymes, and organic acids [11], or lactic acid bacteria and derived bacteriocins [12].

In this context, by-products from fruit and vegetable processing offer an interesting alternative. They are good sources of bioactive compounds and well-recognized for their relevant antimicrobial and antioxidant properties [13,14].

Among fruit and vegetable by-products, pomegranate peel makes up about 40–50% of total fruit weight. This by-product is one of the most abundant wastes discarded during juice, jam, and jelly production [15]. Pomegranate peel contains greater amounts of flavonoids, phenolic acids, tannins, and other compounds than other parts of the fruit [16]. These bioactive compounds have shown important health benefits, including antioxidant and anti-cancer properties [17,18] and remarkable antimicrobial activity against pathogenic and spoilage bacteria [19–23]. Various recent applications have shown how pomegranate peel extract can be used to develop active coatings or bio-based films to be applied to fresh food to control microbial proliferation or oxidation phenomena [24–26]. A few studies have reported the application of peel powder loaded in polymeric matrixes as a potential active food packaging [27–29]. However, the literature shows fewer examples dealing with the direct utilization of pomegranate peel powder for food preservation. In this regard, two studies can be cited. The first one is by Incoronato et al. [30] and deals with the development of new pancakes where both the juice and by-products of pomegranate were used as ingredients. Thus, the addition of by-products to the pancake formulation not only increased its nutritional content but also promoted the extension of its shelf life. Another successful example of pomegranate peel utilization was proposed by Panza et al. [31], who studied how to use pomegranate peel powder as a breading for cod sticks. This research also demonstrated that the use of pomegranate by-products could be a sustainable way to reduce the environmental impacts and costs associated with by-product disposal, with the added advantages of great product quality and increased shelf life.

To raise awareness about complete by-product recycling, their potential applications to food need to be better investigated [32]. Therefore, this study aims to support new advances in the application of pomegranate peel to fresh-cut fruit salad. For the study, the effectiveness of peel powder, at two different concentrations, on the quality decay of a mix of fresh-cut nectarine and pineapple in fructose syrup stored at 4 °C was assessed for 4 weeks. Both the microbiological and sensory qualities were investigated to demonstrate the efficacy of peel by-products on salad quality.

2. Materials and Methods

2.1. Raw Materials and Pomegranate Peel Powder

The pomegranates (*Punica granatum*, cv. Wonderful) were kindly provided by a local horticultural association (A.P.O. Foggia, Italy). The fruits were washed in tap water to remove dust and impurities then dipped for 1 min in chlorinated water (20 mL/L), rinsed to remove chlorine residues, and air dried. The various parts of the fruit were separated manually (peel and arils). The pomegranate peel was cut into small pieces using a sharp knife and dried at 38 °C for 48 h in a dryer (Melchioni-Babele, Milan, Italy). The dried pomegranate peel was finely ground using a laboratory blender, then sieved to obtain a fine powder (500 μm), which was stored in plastic bags at 4 °C and protected from light until its use.

2.2. Fruit Salad Preparation

The nectarine (*Prunus persica*) and pineapple (*Ananas sativus*) were purchased in a local market (Foggia, Italy). The fruits were washed for 1 min in chlorinated water (20 mL/L), rinsed in tap water to remove chlorine residues, and then air dried. The fruits were manually peeled and cut into cubes (1 × 1 cm^2) with a sharp knife. The freshly cut fruits were equally portioned (50 g) into 100 mL polypropylene containers, covered with 70 mL of 25% fructose syrup, then closed using a screw cap in air. Before portioning the fruit and syrup, 2.5% and 5% (*w*/*v*) of pomegranate peel powder were placed at the bottom of each container for the two active samples. These concentrations were found to be the most effective in preliminary analyses carried out by in vitro tests on generic foodborne microorganisms (two species of *Pseudomonas* spp. isolated from spoiled food, identified as *P. fluorescens* and *P. putida*). The control sample (Ctrl) consists of sole fruit salad and fructose syrup without any peel powder added. All of the samples were stored at 5 °C for 28 days. The microbiological and sensory qualities as well as pH were monitored during storage.

2.3. Microbiological Analysis

Under sterile conditions, 20 g of fruit salad was homogenized with a saline solution (0.9% NaCl) (Sigma, Milan, Italy). Decimal dilutions of the homogenate sample were made using the same diluent and plated on selective media to determine the specific microbial groups. Lactic acid bacteria were plated into de Man Rogosa and Sharpe (MRS) agar supplemented with cycloheximide (0.17 g/L) (Sigma, Milan, Italy) and incubated under anaerobic conditions at 37 °C for 48 h. Plate Count Agar (PCA) was used to enumerate the total mesophilic bacterial count incubated at 30 °C for 48 h, and the total psychrotrophic bacteria were incubated for 10 days at 4 °C. Yeasts and molds were determined in Sabouraud Dextrose Agar (SAB), supplemented with chloramphenicol (0.1 g/L) with incubation at 25 °C for 48 h and 5 days, respectively. Violet Red Bile Glucose Agar (VRBGA) incubated at 37 °C for 24 h was instead used for *Enterobacteriaceae*. All of the cultured media and supplements were obtained from Oxoid (Milan, Italy). The analyses were performed in duplicate on different samples, and the results were expressed as log colony-forming units/gram of fruit salad (CFU/g).

2.4. pH Determination

The pH levels of both the homogenized fruit salad and fructose syrup were measured. The measure was performed twice on two different samples by using a pH-meter after the appropriate calibration (Crison, Barcelona, Spain).

2.5. Sensory Evaluation

Seven trained judges, researchers from the University of Foggia, evaluated the sensory quality of the different fruit salad samples. They were already familiar with fresh-cut fruit before this study. However, a brief training section was also carried out to define the sensory attributes to be considered. According to the approach also available in the literature, odor, appearance, flavor, and texture were selected as sensory parameters, and a scale ranging from 1 to 5 (1 = dislike extremely; 2 = dislike moderately; 3 = neither like or dislike; 4 = like moderately; and 5 = like extremely) was used for the evaluation [33]. During the sensory analysis, the control and active fruit salads were differently coded and presented in random order to the panelists. Individually, they expressed their degree of appreciation for each attribute, and finally, using the same scale, they were also asked to judge the overall quality of each salad sample.

2.6. Statistical Analysis

Tests were carried out on duplicate batches. Experimental data are the average of two replicates. The results are presented as mean ± Standard Deviation (SD) and graphically reported. A statistical significance was determined by a one-way analysis of variance (ANOVA). Duncan's multiple range test, with the option of homogeneous groups ($p \leq 0.05$), was performed to determine significant differences among fruit salad samples. For this, STATISTICA 7.1 for Windows (StatSoft, Inc, Tulsa, OK, USA) was used.

3. Results and Discussion

3.1. Microbial Quality Decay during Refrigerated Storage

Fresh-cut fruit has a high susceptibility to microbial spoilage. High levels of carbohydrates and water and low pH values make the environment optimal for the growth of mesophilic, psychrotrophic, and lactic acid bacteria, in addition to yeasts and molds [12,34]. Therefore, to assess the effect of different concentrations of pomegranate peel powder (2.5 and 5%) on the microbial quality of fruit salad, the viable cell concentration of the main spoilage groups was monitored. The two percentages of peel used in this study were chosen based on preliminary in vitro antimicrobial tests (data not shown). As reported in the M&M Section 2, the evolution of microbial growth in fruit salad stored under refrigerated conditions (5 °C) was monitored for 28 days. During the storage period, statistically significant differences were observed between fruit salad with and without pomegranate peel powder, with the active samples being less contaminated.

In particular, the evolution of total mesophilic and psychrotrophic bacterial counts is shown in Figure 1ab. As can be seen, for the total mesophilic bacteria (Figure 1a), the initial microbial concentration of both the control and active fruit salads was around 3.80 log CFU/g. During the first 8 days, the Ctrl and active samples maintained, more or less, the initial microbial concentration; in the following days, a marked difference appeared between them. Specifically, the Ctrl sample showed a significant microbial increase up to 8.27 log CFU/g after 28 days, whereas, both salads with peel powder maintained the same microbial count for more than 2 weeks. After this long lag phase, the bacteria gradually grew; however, the increase in the viable cell concentration of active samples was less pronounced than that recorded for the control sample. The control salad reached 8 log CFU/g, whereas both active systems remained around 7 log CFU/g. Most probably, the pomegranate peel powder, rich in active compounds, inhibited microbial growth during the first stage of storage, and subsequently, microorganisms, accustomed to the conditions, began to grow [17,35]. Sun et al. [23] suggested that the antimicrobial activity of pomegranate peels is related to the combined effect of polyphenols, sterols, and pentacyclic triterpenoid compounds.

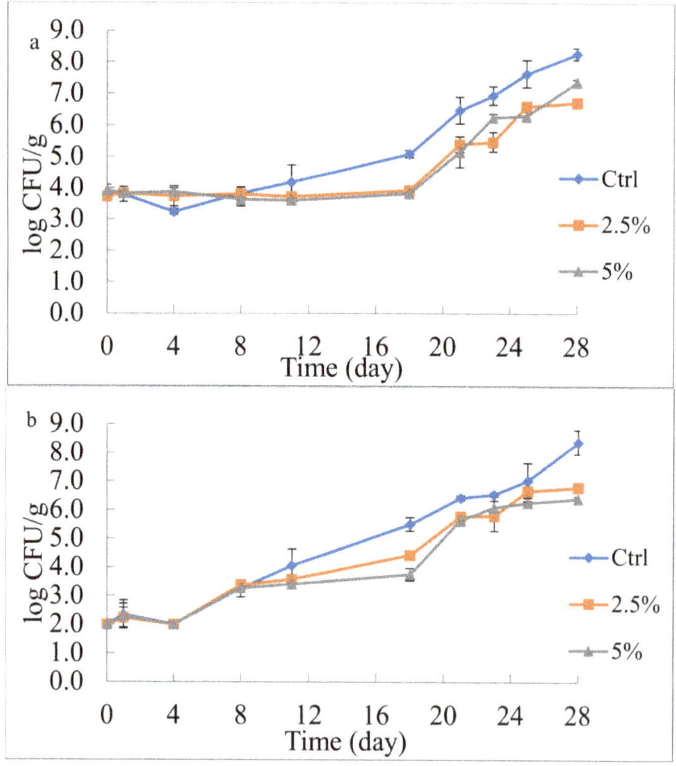

Figure 1. Evolution of total mesophilic (**a**) and psychrotrophic (**b**) bacteria in fruit salad during 28 days of refrigerated storage (5 °C). Ctrl: fruit salad without pomegranate peel powder; 2.5%: fruit salad with 2.5% (*w*/*v*) pomegranate peel powder; 5%: fruit salad with 5% (*w*/*v*) pomegranate peel powder.

For the total psychrotrophic bacteria, a trend similar to the total mesophilic count was observed (Figure 1b). During the exponential phase, and also at the stationary phase, in the active sample, the microbial load was lower than that observed in the control sample. In fact, also in this case, control salad reached 8 log CFU/g, whereas, the active samples remained between 6 and 7 log CFU/g after the 4 weeks of observation. These antimicrobial effects on different microbial groups are not surprising because it is well recognized from studies reported in the literature that the bioactive substances contained in pomegranate peel can inhibit growth of different microbial species [15,21,27].

Looking at recorded data, product acceptability in terms of microbial quality can be defined. According to the French Regulation, fresh-cut fruit remained acceptable until the total count of mesophilic and psychrotrophic bacteria reaches 5×10^7 CFU/g [36]. Considering the results of the control and active systems reported in Figure 1 1a,b, it can be inferred that the control salad remained acceptable for about 24 days, whereas, both salads with pomegranate peel powder were found below the microbial threshold for the entire observation period (28 days). Therefore, according to these experimental findings, the lowest concentration of pomegranate by-product is enough to assure a longer microbial stability of fresh-cut nectarine and pineapple mix than the control salad.

While slight differences in pH values between the control and active systems were found, no differences in pH were recorded between the fruit and its fructose syrup (data not shown). To give a more precise idea about the product pH, after 28 days of storage the pH values observed for the active samples were about 3.72 and 3.68 for the 2.5 and 5% samples, respectively, whereas, a pH of about 4.46 was recorded for the control sample.

Regarding the effects of the by-products on the other monitored spoilage groups, it is worth noting that the peel powder added to fruit salad significantly affected yeast growth, as shown in Figure 2. Although the initial microbial concentration of the active samples was slightly lower than the control, the trend of the three investigated samples was similar during the first 8 days. After this period, yeast growth rate changed. In particular, a more rapid increase was found in the control system and delayed kinetics were recorded for both active packages. A final count of 6.22 log CFU/g was measured in the Ctrl sample, whereas, 5.80 and 4.79 log CFU/g were found in the active samples with 2.5 and 5% pomegranate peel powder, respectively. The fruit salad with the highest concentration of pomegranate peel was the least contaminated at the end of the storage period, thus confirming the antimicrobial activity of this by-product against fungal spoiling [16]. Gull et al. [26] also observed that chitosan coating enriched with pomegranate peel extract was effective in protecting apricot from yeast spoiling when stored for 30 days.

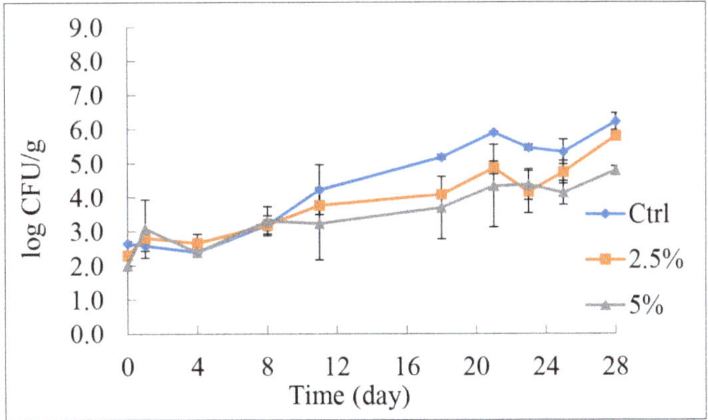

Figure 2. Evolution of yeasts in fruit salad during 28 days of refrigerated storage (5 °C). Ctrl: fruit salad without pomegranate peel powder; 2.5%: fruit salad with 2.5% (w/v) pomegranate peel powder; 5%: fruit salad with 5% (w/v) pomegranate peel powder.

In relation to lactic acid bacteria (LAB), the effects of the peel powder were very marked. It is also striking to observe that between the two concentrations of by-product, the highest one was the most effective (Figure 3). In particular, in the control sample, LAB remained low for one week and then increased up to about 5 log CFU/g. In both active salads, LAB had a long lag phase with a very small microbial load for more than two weeks. The cells then grew; however, the viable cell concentration reached the value of about 5 log CFU/g in the sample with 2.5% peel powder and about 4 log CFU/g in the sample with the highest peel powder concentration.

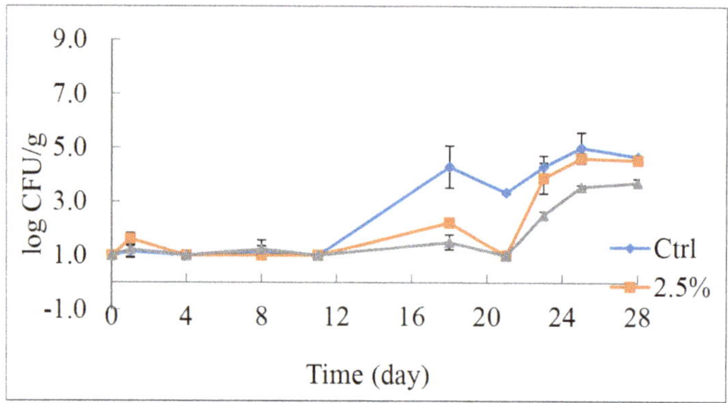

Figure 3. Evolution of lactic acid bacteria in fruit salad during 28 days of refrigerated storage (5 °C). Ctrl: fruit salad without pomegranate peel powder; 2.5%: fruit salad with 2.5% (*w*/*v*) pomegranate peel powder; 5%: fruit salad with 5% (*w*/*v*) pomegranate peel powder.

Polyphenols contained in the active powder can justify the observed antimicrobial activity [35,37]. One of the factors that can influence the efficacy of these compounds against different microbial and fungal groups is the position of the hydroxyl groups (OH) in the aromatic ring of polyphenols. Hydroxyl groups can interact with microbial cell membranes to make them more permeable and cause the loss of cellular components, as well as damage microbial metabolic processes [35]. A representation of the complex mechanisms involved in the antimicrobial effects of polyphenols is also provided in Figure 4.

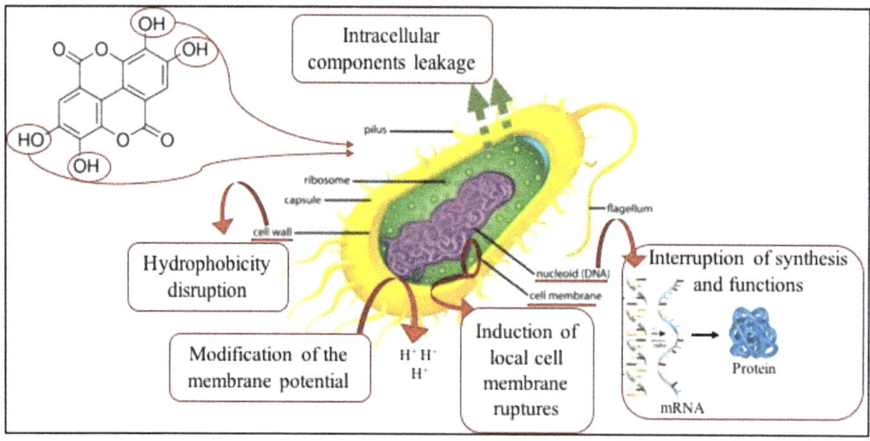

Figure 4. Mechanism of action of polyphenols on a microbial cell.

Regarding mold proliferation, in all packages a concentration of about 3 log CFU/g was measured, without any statistically significant differences among samples (data not shown). With high probability, the washing of fruit in chlorinated water before cutting and the refrigerated storage conditions controlled the mold proliferation [10].

No significant differences were also found in the viable cell concentration of *Enterobacteriaceae*, which recorded a final concentration of about 4.5 log CFU/g in all investigated samples (data not shown). This finding about *Enterobacteria* means that general hygienic conditions were adopted during production and processing.

3.2. Sensory Quality Decay during Refrigerated Storage

With regard to the effects of pomegranate peel on product acceptability, the changes in sensory quality during storage are reported in Figure 5. As one would expect, during the entire storage period, a gradual decrease in the sensory quality was found in both the Ctrl and active samples [33]. As can be inferred from the data shown in the figure, fruit salads, regardless of the type of sample, remained acceptable for around 3 weeks. Then, defects appeared that made the product disagreeable to panelists. Table 1 lists the specific sensory attribute scores as assessed by the panel. The data highlight that the main attribute responsible for product sensory deterioration is the odor; texture and general appearance also contributed to product rejection. This finding is not surprising because microbial and fungal proliferation, water loss, and enzymatic reactions occurring during fruit storage generally also cause changes in the product quality, primarily in terms of odor and color [2,38]. Loss of firmness can be correlated with tissue degradation [4,5]. Therefore, the trends shown in Figure 4 for fruit salads' overall quality effectively reflect the decrease in odor, appearance, and texture that occurred in all of the samples.

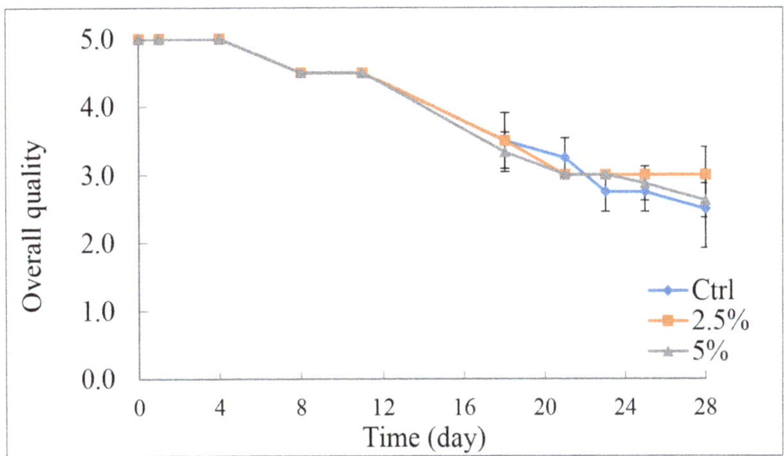

Figure 5. Overall quality of fruit salad during 28 days of refrigerated storage (5°C). Ctrl: fruit salad without pomegranate peel powder; 2.5%: fruit salad with 2.5% (*w*/*v*) pomegranate peel powder; 5%: fruit salad with 5% (*w*/*v*) pomegranate peel powder.

When comparing the fruit salad sensory quality of the investigated samples, it appears that pomegranate peel powder did not worsen the main sensory fruit salad attributes. On the contrary, the panelists appreciated a slight herbaceous smell and a marked fruity aroma in the active samples that were stored for about 20 days.

The enhancement in sensory quality was also recorded when pomegranate peel powder was added to pancake formulations [30] and when the same by-product was adopted as a breading for fresh cod sticks [31].

Table 1. Sensory attributes of active and non-active fresh fruit salad during 28 days of storage at 5 °C.

Samples		Storage Time (Days)									
		0	1	4	8	11	18	21	23	25	28
Appearance	Control	5.0 ± 0.0 [a]	5.0 ± 0.0 [a]	5.0 ± 0.0 [a]	4.5 ± 0.0 [a]	4.5 ± 0.0 [b]	3.5 ± 0.5 [a]	3.4 ± 0.3 [b]	3.3 ± 0.3 [a]	3.4 ± 0.3 [b]	2.9 ± 0.3 [a]
	2.5%	5.0 ± 0.0 [a]	5.0 ± 0.0 [a]	5.0 ± 0.0 [a]	4.5 ± 0.5 [a]	3.9 ± 0.3 [a]	3.2 ± 0.3 [a]	3.0 ± 0.0 [a]	3.0 ± 0.0 [a]	3.0 ± 0.0 [a,b]	2.9 ± 0.3 [a]
	5%	5.0 ± 0.0 [a]	5.0 ± 0.0 [a]	5.0 ± 0.0 [a]	4.2 ± 0.3 [a]	3.9 ± 0.3 [a]	3.2 ± 0.3 [a]	3.0 ± 0.0 [a]	3.0 ± 0.0 [a]	2.9 ± 0.3 [a]	3.0 ± 0.0 [a]
Odor	Control	5.0 ± 0.0 [a]	5.0 ± 0.0 [a]	5.0 ± 0.0 [a]	4.5 ± 0.0 [a]	4.0 ± 0.0 [a]	3.8 ± 0.3 [a]	3.2 ± 0.3 [a]	2.8 ± 0.3 [a]	2.9 ± 0.3 [a]	2.3 ± 0.5 [a]
	2.5%	5.0 ± 0.0 [a]	5.0 ± 0.0 [a]	5.0 ± 0.0 [a]	5.0 ± 0.0 [b]	4.5 ± 0.0 [b]	3.9 ± 0.3 [a]	2.9 ± 0.3 [a]	2.8 ± 0.3 [a]	3.0 ± 0.0 [a]	3.0 ± 0.3 [a]
	5%	5.0 ± 0.0 [a]	5.0 ± 0.0 [a]	5.0 ± 0.0 [a]	4.5 ± 0.0 [a]	4.5 ± 0.0 [b]	3.5 ± 0.0 [a]	2.9 ± 0.3 [a]	2.9 ± 0.3 [a]	2.9 ± 0.3 [a]	2.7 ± 0.3 [a]
Texture	Control	5.0 ± 0.0 [a]	5.0 ± 0.0 [a]	5.0 ± 0.0 [a]	4.5 ± 0.0 [a]	4.5 ± 0.0 [a]	3.5 ± 0.0 [a]	3.3 ± 0.3 [a]	3.5 ± 0.0 [a]	2.9 ± 0.3 [a]	2.9 ± 0.3 [a]
	2.5%	5.0 ± 0.0 [a]	5.0 ± 0.0 [a]	5.0 ± 0.0 [a]	4.5 ± 0.0 [a]	4.5 ± 0.0 [a]	3.9 ± 0.3 [a]	3.5 ± 0.0 [a]	3.3 ± 0.3 [a]	3.3 ± 0.3 [a]	3.0 ± 0.3 [a]
	5%	5.0 ± 0.0 [a]	5.0 ± 0.0 [a]	5.0 ± 0.0 [a]	4.5 ± 0.0 [a]	4.5 ± 0.0 [a]	3.9 ± 0.3 [a]	3.5 ± 0.0 [a]	3.4 ± 0.3 [a]	3.2 ± 0.3 [a]	3.0 ± 0.0 [a]

For each sensory attribute, data (±SD, $n = 2$) marked with different superscript letters ([a,b]) in each column are significantly different ($p < 0.05$). Control: fruit salad without pomegranate peel powder; 2.5%: fruit salad with 2.5% (w/v) pomegranate peel powder; 5%: fruit salad with 5% (w/v) pomegranate peel powder.

In the current study, the overall quality decay of the investigated samples reported in Figure 5 highlights that the degradation of the control sample after 3 weeks was very rapid. On the contrary, in the active samples the sensory quality decay was slower. The reasons for the abovementioned difference among samples are strictly linked to the fact that the control sample becomes unacceptable because of the degradation of the whole product, whereas the active samples were mainly refused for the turbidity of the syrup in the containers, which negatively affected the visual quality of the fruit salad. This drawback represents one of the main problems faced when by-products are applied to food and necessitates further research to find valid solutions to boost by-product recycling [32]. The inclusion of peel powder in either an edible or biodegradable polymeric matrix could be a valid strategy [14].

4. Conclusions

Fresh-cut fruits, although providing health benefits and convenience to consumers, are highly perishable. This study proves that traditional non-edible parts of pomegranate fruit, such as the peel, can be used to slow down the detrimental phenomena responsible for fruit salad unacceptability. In particular, the current research investigated the effects of 2.5 and 5% pomegranate peel powder placed on the bottom of a fruit salad container filled with fructose syrup. The investigated by-product exerted good antimicrobial and antifungal activity; in fact, it prolonged the lag phase and reduced the cell viable concentration at the stationary phase of total mesophilic and psychrotrophic bacteria, yeasts, and lactic acid bacteria, primarily when the highest peel powder concentrations were used. Consequently, it is possible to conclude that fruit salads with 2.5 and 5% of pomegranate peel powder were less contaminated than the control sample. The effects of peel powder are further prized if the sensory quality of fruit is considered. In fact, the investigated by-product did not compromise fruit acceptability; moreover, similar sensory quality decay kinetics were recorded in the control and active samples. Further studies are needed to solve the problem of suspended powder in fructose syrup because this flaw can make the product unattractive. Therefore, considering the positive impacts by-product valorization has on both the economy and the environment, we have another study in progress to optimize the addition of this valuable peel powder to make the final product more appealing.

Author Contributions: Conceptualization, M.A.D.N. and A.C.; methodology, M.A.D.N., A.C.; formal analysis, A.L.I. and V.L.; data curation, M.A.D.N., A.C.; writing—original draft preparation, V.L. and A.C.; writing—review and editing, M.A.D.N.; A.C.; supervision M.A.D.N. and A.C. All authors have read and agreed to the published version of the manuscript.

Funding: This research received no external funding.

Data Availability Statement: The raw data will be made available upon request.

Conflicts of Interest: The authors declare no conflict of interest.

References

1. Ruiz Rodríguez, L.G.; Gasga, V.M.Z.; Pescuma, M.; Van Nieuwenhove, C.; Mozzi, F.; Sánchez Burgos, J.A. Fruits and fruit by-products as sources of bioactive compounds. Benefits and trends of lactic acid fermentation in the development of novel fruit-based functional beverages. *Food Res. Int.* **2021**, *140*, 109854. [CrossRef] [PubMed]
2. Barrett, D.M.; Beaulieu, J.C.; Shewfelt, R. Color, flavor, texture, and nutritional quality of fresh-cut fruits and vegetables: Desirable levels, instrumental and sensory measurement, and the effects of processing. *Crit. Rev. Food Sci. Nutr.* **2010**, *50*, 369–389. [CrossRef] [PubMed]
3. Gil, M.I.; Aguayo, E.; Kader, A.A. Quality changes and nutrient retention in fresh-cut versus whole fruits during storage. *J. Agric. Food Chem.* **2006**, *54*, 4284–4296. [CrossRef] [PubMed]
4. Finnegan, E.; O'Beirne, D. Characterising deterioration patterns in fresh-cut fruit using principal component analysis. II: Effects of ripeness stage, seasonality, processing and packaging. *Postharvest Biol. Technol.* **2015**, *100*, 91–98. [CrossRef]
5. Finnegan, E.; O'Beirne, D. Characterising and tracking deterioration patterns of fresh-cut fruit using principal component analysis—Part I. *Postharvest Biol. Technol.* **2015**, *100*, 73–80. [CrossRef]
6. Li, X.; Long, Q.; Gao, F.; Jin, P.; Zheng, Y. Effect of cutting styles on quality and antioxidant activity in fresh-cut pitaya fruit. *Postharvest Biol. Technol.* **2017**, *124*, 1–7. [CrossRef]

7. Zheng, H.; Liu, W.; Liu, S.; Liu, C.; Zheng, L. Effects of melatonin on the enzymatic browning and nutritional quality of fresh-cut pear fruit. *Food Chem.* **2019**, *299*, 125116. [CrossRef]
8. Cortellino, G.; Gobbi, S.; Bianchi, G.; Rizzolo, A. Modified atmosphere packaging for shelf life extension of fresh-cut apples. *Trends Food Sci. Technol.* **2015**, *46*, 320–330. [CrossRef]
9. Maringgal, B.; Hashim, N.; Tawakkal, I.S.M.A.; Mohamed, M.T.M. Recent advance in edible coating and its effect on fresh/fresh-cut fruits quality. *Trends Food Sci. Technol.* **2020**, *96*, 253–267. [CrossRef]
10. Meireles, A.C.; Giaouris, E.; Simões, M. Alternative disinfection methods to chlorine for use in the fresh-cut industry. *Food Res. Int.* **2016**, *82*, 71–85. [CrossRef]
11. Mastromatteo, M.; Conte, A.; Del Nobile, M.A. Combined use of modified atmosphere packaging and natural compounds for food preservation. *Food Eng. Rev.* **2010**, *2*, 28–38. [CrossRef]
12. Agriopoulou, S.; Stamatelopoulou, E.; Sachadyn-Król, M.; Varzakas, T. Lactic acid bacteria as antibacterial agents to extend the shelf life of fresh and minimally processed fruits and vegetables: Quality and safety aspects. *Microorganisms* **2020**, *8*, 952. [CrossRef]
13. Sagar, N.A.; Pareek, S.; Sharma, S.; Yahia, E.M.; Lobo, M.G. Fruit and vegetable waste: Bioactive compounds, their extraction, and possible utilization. *Compr. Rev. Food Sci. Food Saf.* **2018**, *17*, 512–531. [CrossRef]
14. Dilucia, F.; Lacivita, V.; Conte, A.; Del Nobile, M.A. Review: Sustainable use of fruit and vegetable by-products to enhance food packaging performance. *Foods* **2020**, *9*, 857. [CrossRef]
15. Gullon, B.; Pintado, M.E.; Pérez-Àlvarez, J.A.; Viuda-Martos, M. Assessment of polyphenolic profile and antibacterial activity of pomegranate peel (*Punica granatum*) flour obtained from co-product of juice extraction. *Food Control* **2016**, *59*, 94–98. [CrossRef]
16. Akhatar, S.; Ismail, T.; Fraternale, D.; Sestili, P. Review: Pomegranate peel and peel extracts: Chemistry and food features. *Food Chem.* **2015**, *174*, 417–425. [CrossRef]
17. Fischer, U.A.; Carle, R.; Kammerer, D.R. Identification and quantification of phenolic compounds from pomegranate (*Punica granatum* L.) peel, mesocarp, aril and differently produced juices by HPLC-DAD–ESI/MSn. *Food Chem.* **2011**, *127*, 807–821. [CrossRef]
18. Mushtaq, M.; Sultana, B.; Anwar, F.; Adnan, A.; Rizvi, S.S. Enzyme-assisted supercritical fluid extraction of phenolic antioxidants from pomegranate peel. *J. Supercrit. Fluids* **2015**, *104*, 122–131. [CrossRef]
19. Molva, Ç.; Baysal, A.H. Evaluation of bioactivity of pomegranate fruit extract against Alicyclobacillus acidoterrestris DSM 3922 vegetative cells and spores in apple juice. *LWT Food Sci. Technol.* **2015**, *62*, 989–995. [CrossRef]
20. Nicosia, M.G.L.D.; Pangallo, S.; Raphael, G.; Romeo, F.V.; Strano, M.C.; Rapisarda, P.; Schena, L. Control of postharvest fungal rots on citrus fruit and sweet cherries using a pomegranate peel extract. *Postharvest Biol. Technol.* **2016**, *114*, 54–61. [CrossRef]
21. Kharchoufi, S.; Licciardello, F.; Siracusa, L.; Muratore, G.; Hamdi, M.; Restuccia, C. Antimicrobial and antioxidant features of 'Gabsi' pomegranate peel extracts. *Ind. Crops Prod.* **2018**, *111*, 345–352. [CrossRef]
22. Kandylis, P.; Kokkinomagoulos, E. Review: Food applications and potential health benefits of pomegranate and its derivatives. *Foods* **2020**, *9*, 122. [CrossRef]
23. Sun, S.; Huang, S.; Shi, Y.; Shao, Y.; Qiu, J.; Sedjoah, R.-C.A.-A.; Yan, Z.; Ding, L.; Zou, D.; Xin, Z. Extraction, isolation, characterization and antimicrobial activities of non-extractable polyphenols from pomegranate peel. *Food Chem.* **2021**, *351*, 129232. [CrossRef]
24. Mushtaq, M.; Gani, A.; Gani, A.; Punoo, H.A.; Masoodi, F.A. Use of pomegranate peel extract incorporated zein film with improved properties for prolonged shelf life of fresh Himalayan cheese (Kalari/kradi). *Innov. Food Sci. Emerg. Technol.* **2018**, *48*, 25–32. [CrossRef]
25. Nair, M.S.; Saxena, A.; Kaur, C. Effect of chitosan and alginate-based coatings enriched with pomegranate peel extract to extend the postharvest quality of guava (*Psidium guajava* L.). *Food Chem.* **2018**, *240*, 245–252. [CrossRef]
26. Gull, A.; Hat, N.; Wani, S.M.; Masoodi, F.A.; Amin, T.; Ganai, S.A. Shelf life extension of apricot fruit by application of nanochitosan emulsion coatings containing pomegranate peel extract. *Food Chem.* **2021**, *349*, 129149. [CrossRef] [PubMed]
27. Hanani, Z.A.N.; Husna, A.B.A.; Syahida, S.N.; Nor-Khaizura, M.A.B.; Jamilah, B. Effect of different fruit peels on the functional properties of gelatin/polyethylene bilayer films for active packaging. *Food Packag. Shelf Life* **2018**, *18*, 201–211. [CrossRef]
28. Hanani, Z.A.N.; Yee, F.C.; Nor-Khaizura, M.A.R. Effect of pomegranate (Punica granatum L.) peel powder on the antioxidant and antimicrobial properties of fish gelatin films as active packaging. *Food Hydrocoll.* **2019**, *89*, 253–259. [CrossRef]
29. Ali, A.; Chen, Y.; Liu, H.; Yu, L.; Baloch, Z.; Khalid, S.; Zhu, J.; Chen, L. Starch-based antimicrobial films functionalized by pomegranate peel. *Int. J. Biol. Macromol.* **2019**, *129*, 1120–1126. [CrossRef]
30. Incoronato, A.L.; Cedola, A.; Conte, A.; Del Nobile, M.A. Juice and by-products from pomegranate to enrich pancake: Characterization and shelf-life evaluation. *Int. J. Food Sci. Technol.* **2021**, *56*, 2886–2894. [CrossRef]
31. Panza, O.; Conte, A.; Del Nobile, M.A. Pomegranate by-products as natural preservative to prolong the shelf life of breaded cod stick. *Molecules* **2021**, *26*, 2385. [CrossRef] [PubMed]
32. Majerska, J.; Michalska, A.; Figiel, A. A review of new directions in managing fruit and vegetable processing by-products. *Trends Food Sci. Technol.* **2019**, *88*, 207–219. [CrossRef]
33. Costa, C.; Conte, A.; Buonocore, G.G.; Del Nobile, M.A. Antimicrobial silver montmorillonite nanoparticles to prolong the shelf life of fresh fruit salad. *Int. J. Food Microbiol.* **2011**, *148*, 164–167. [CrossRef] [PubMed]

34. Qadri, O.S.; Yousuf, B.; Srivastava, A.K. Fresh-cut fruits and vegetables: Critical factors influencing microbiology and novel approaches to prevent microbial risks—A review. *Cogent Food Agric.* **2015**, *1*, 1121606. [CrossRef]
35. Alexandre, E.M.C.; Silva, S.; Santos, S.A.O.; Silvestre, A.J.D.; Duarte, M.F.; Saraiva, J.A.; Pintado, M. Antimicrobial activity of pomegranate peel extracts performed by high pressure and enzymatic assisted extraction. *Food Res. Int.* **2019**, *115*, 167–176. [CrossRef]
36. Ministere de l'Economie des Finances et du Budget. Marché consommation, Produits vegetaux prets a l'emploi dits de la 'IVemme Gamme': Guide de bonnes pratique hygieniques. *J. Officiel de la Republique Francaise* **1988**, *1621*, 1–29.
37. Andrade, M.A.; Lima, V.; Sanches Silva, A.; Vilarinho, F.; Castilho, M.C.; Khwaldia, K.; Ramos, F. Pomegranate and grape by-products and their active compounds: Are they a valuable source for food applications? *Trends Food Sci. Technol.* **2019**, *86*, 68–84. [CrossRef]
38. Yousuf, B.; Deshi, V.; Ozturk, B.; Siddiqui, M.W. *Fresh-Cut Fruits and Vegetables: Quality Issues and Safety Concerns*; Fresh-Cut Fruits and Vegetables; Siddiqui, M.W., Ed.; Academic Press: Cambridge, MA, USA, 2020; pp. 1–15.

 MDPI

Article

Packaging Solutions to Extend the Shelf Life of Green Asparagus (*Asparagus officinalis* L.) 'Vegalim'

Stefania Toscano [1], Valeria Rizzo [1,*], Fabio Licciardello [2], Daniela Romano [1] and Giuseppe Muratore [1]

1 Department of Agriculture, Food and Environment (Di3A), University of Catania, Via Santa Sofia, 100, 95123 Catania, Italy; stefania.toscano@unict.it (S.T.); dromano@unict.it (D.R.); giuseppe.muratore@unict.it (G.M.)
2 Department of Life Sciences, University of Modena and Reggio Emilia, Via Amendola 2, 42122 Reggio Emilia, Italy; fabio.licciardello@unimore.it
* Correspondence: vrizzo@unict.it

Abstract: The aim of the study was to assess, through a comparative shelf-life test, the suitability of two packaging materials, namely macro-perforated polypropylene (PP MA) and micro-perforated coextruded polypropylene (PP C), for the quality preservation of green asparagus (*Asparagus officinalis* L. 'Vegalim'). Quality of spears was evaluated during 30 days at refrigerated storage by monitoring chemical, physical, and enzymatic parameters as well as sensory descriptors. PP C kept headspace composition close to suggested values for fresh green asparagus. Total color difference increased during the storage and it was highly correlated with chlorophyll-*a* and carotenoids, however, sensory color perception did not change significantly until 22 days of storage. PP C maintained ascorbic acid concentrations close to the initial levels, limited total phenolic compound loss to 24% (45% in PP MA), determined an increase of 72% in fiber content and small changes in lignin value; enzymatic changes were significantly inhibited. Significant sensorial differences were detected after 22 days of storage, with PP C performing better than PP MA. PP C film was confirmed as the best choice, limiting weight loss and maintaining a fresh-like appearance during 30 days of storage, thus allowing an extension in postharvest life.

Keywords: asparagus; enzyme activity; lignin; fiber; weight loss; color; polypropylene film

Citation: Toscano, S.; Rizzo, V.; Licciardello, F.; Romano, D.; Muratore, G. Packaging Solutions to Extend the Shelf Life of Green Asparagus (*Asparagus officinalis* L.) 'Vegalim'. *Foods* **2021**, *10*, 478. https://doi.org/10.3390/foods10020478

Academic Editor: Benu P. Adhikari

Received: 2 February 2021
Accepted: 19 February 2021
Published: 22 February 2021

1. Introduction

Recently, thanks to its high dietary value and the presence of vitamin C, amino acids, sterols, and fiber, as well as bioactive compounds, such as flavonoids and other polyphenolic compounds, including quercetin and cinnamic acid [1,2], asparagus (*Asparagus officinalis* L.) has been increasingly consumed [3]. Therefore, the need has arisen for effective preservation strategies that might retard quality loss, thus allowing long-distance distribution and extended shelf life. Asparagus is a highly perishable vegetable, both for its delicate structure and for its proneness to physiological alterations [4]. During storage, asparagus spear quality is reduced due to toughening, loss of water, and changes the levels of in ascorbic acid, carbohydrates, protein, and amino acids [5].

The short shelf life of asparagus is mainly due to its high respiration rate of about 60 mg CO_2/kg × h at 5 °C [6], which continues after harvesting.

'Vegalim' is a male hybrid that is ideal for green asparagus production in regions with a warm or Mediterranean climate. The main characteristics are a medium-early production with high yields and excellent close tips, even in hot or warm climates. Fertile and well drained soils are preferred. This variety is bred in the Netherlands and it is suggested for green asparagus production under warm conditions, so the spear opening is delayed as temperature rises [7]. Nevertheless, the cultivar shows good adaptability to winter Mediterranean climate, so the production normally starts in January. It can also be adapted to the production of white asparagus, and spears are characterized by high weight and uniform size.

Fresh vegetables should ideally retain freshness during the distribution life, and be reasonably defect-free—thus, suitable packaging solutions should be designed to maintain the highest quality possible for the longest time. After harvest, asparagus spears are subjected to different changes, which result in the loss of quality, including the withering of spears, changes in sugar and organic acid content, chlorophyll degradation, and lignification of pericyclic fibers that results in toughening of tissue [8,9]. The above undesirable effects can be reduced by quick cooling upon harvesting, refrigeration (below 5 °C for long term storage), and storage in a modified atmosphere [10]. Sensory characteristics are used by customers as quality standards in vegetable foods; however, the food industry needs simpler methods to visualize the sensory quality achieved by producers rather than the intensities of the descriptors [11].

The degree of lignification is one of the principal factors in determining the quality of green asparagus. Lignin biosynthesis is associated with phenylalanine ammonia-lyase (PAL) and peroxidase (POD) activity in plant tissues. Being able to inhibit or regulate the activity of POD and PAL enzymes could actually improve the quality of green asparagus [12].

Senescence is also associated with the defense system that triggers the response of antioxidant enzymes and antioxidant substances naturally present in the vegetables. Antioxidant enzymes (superoxide dismutase (SOD), catalase (CAT) and peroxidase (POD), for example) are of fundamental importance in the radical detoxification process in plant tissues [13,14].

Packaging plays a key role in quality preservation of foods. With regards to fresh horticultural products, packaging may limit water vapor transmission and regulate gas exchange. In turn, these events result in the reduction of weight loss and withering degree and in the slow-down of respiration. Since vegetable senescence is associated with the extent of metabolic activities, the slow-down of respiration will result in prolonged quality maintenance and retention of bioactive components [15]. The selection of suitable packaging materials is crucial, since inappropriate packaging properties may reduce, rather than extend, the product shelf life, by determining unsuitable headspace conditions resulting in a metabolic imbalance [16].

Overall, the high economic value of this vegetable and its high perishability have raised the need to look for possible strategies to increase its shelf life, since nowadays asparagus spears are often distributed unpacked, in bunches surrounded by paperboard or plastic bands. Therefore, the aim of the current study was to assess the performance of two packaging solutions that could be applied by producers, in terms of potential for shelf-life extension of green asparagus. The study was carried out by monitoring changes in color, nutritional parameters, and sensory properties related to physiological changes, with a specific focus on the enzymatic changes occurring during refrigerated storage.

2. Materials and Methods

2.1. Biological Material

Fresh green asparagus spears (*Asparagus officinalis* L. 'Vegalim') were kindly provided by Asparago Sovrano Consortium (Enna, Italy).

Commercially mature spears of 28.5 ± 1.5 cm were harvested and transported to the Di3A laboratories, then undamaged and straight spears, of about 16–20 mm in diameter and ~22 cm in length with closed bracts and no visible signs of injury were selected for the study.

2.2. Sample Preparation and Packaging Material

Green asparagus ('Vegalim') were divided into 2 homogeneous batches and packaged into polypropylene macro-perforated bags (PP MA) with a piercing density of 7 holes cm^{-2} (Bemis Le Trait sas, Bourg Achard, France) and in micro-perforated coextruded PP bags

(CoralifeSwaf C, Corapack s.r.l., Brenna (CO), Italy), with a row of central holes spaced 3 cm apart and having an oxygen transmission rate of 38,800 cc m^{-2} (PP C). In total, 60 bags (20 × 35 cm) were prepared using a manual sealing bar and filled with 65 ± 10 g of fresh asparagus spears. All tasters were kept under refrigerated conditions at 4 ± 1 °C and 90–95% Relative Humidity (RH) until analyses. Considering the natural decay of the product, quality parameters were monitored every 7 days and performed on three replicates for each batch; reported results were after 0 (packaging day), 8, 15, 22 and 30 days of storage.

2.3. Headspace Gas Composition and Color Determination

The headspace gas composition was measured as reported by Villanueva et al. [3] using a Dansensor Checkpoint (Dansensor, Ringsted, Denmark) and analyzing 10 mL of the package headspace on three replicates.

Color was measured on three spears randomly chosen from two different bags, with a Minolta portable colorimeter (CR-400 Konica Minolta, Inc., Osaka, Japan) using the CIEL*a*b* color parameters, L*, a*, and b*. The equipment was calibrated by a standard white calibration plate and using C as illumination source. Total color difference ΔE* was calculated as reported by Licciardello et al. [17].

2.4. Chemical Analyses

Ascorbic acid was determined according to Association of Official Agricultural Chemists (AOAC) [18]; briefly 10 g of samples were homogenized by Ultraturrax for 2 min in 40 mL extraction solution prepared by dissolving 15 g metaphosphoric acid in 40 mL acetic acid and bringing to a final volume of 500 mL with distilled water; 2 mL 2,6 dichloroindophenol diluted 1:5 was mixed with 0.5 mL extract and the absorbance was monitored at 518 nm by a Shimadzu 1601 UV-Visible spectrophotometer (Shimadzu Corp. Tokyo, Japan). The quantification was obtained by a calibration curve of suitable dilutions of standard ascorbic acid.

Total polyphenols were measured and quantified using the Folin–Ciocalteu spectrophotometric method [19] The absorbance was measured at 740 nm using a Shimadzu 1601 UV-Visible spectrophotometer (Shimadzu Corp. Tokyo, Japan), and the total polyphenol content (TPC) was then determined on the basis of a standard calibration curve generated with known amounts of gallic acid. Data were expressed as g of gallic acid equivalent (GAE) kg^{-1} of fresh weight.

Chlorophylls and carotenoids were determined according to Li and Zhang [20] and Lichtenthaler [21]. Briefly, 2 g of asparagus were homogenized with acetone and analyzed, after centrifuging, at 645 and 662 nm for chlorophyll *a* (Chl-a) and *b* (Chl-b), respectively, and at 470 nm for carotenoids.

Fiber content was determined as reported by Sothornvit and Kiatchanapaibul [22]. Samples were heated in hot water and 50% (*w*/*v*) NaOH; then, fiber was dried and weighed to calculate the percentage of fiber content, while lignin content was determined using the thioacidoglycolysis method, as proposed by Techavuthiporn and Boonyaritthongchai [23]. The absorbance was read at 280 nm using a UV-2401 spectrophotometer (Shimadzu Co., Kyoto, Japan) and the quantification was obtained using standard curves of p-coumaric acid; lignin content was expressed as g kg^{-1} of fresh weight.

All chemical analyses were performed in triplicate. Reagents and solvents were purchased from Sigma-Aldrich (Milan, Italy) and were of analytical or HPLC grade. Bi-distilled water was used throughout these tests.

2.5. Enzyme Analysis and MDA Content

For antioxidant enzyme assays, 0.4 g of asparagus were ground to a fine powder with liquid nitrogen and was extracted with 6 mL of extraction buffer (0.1 mol L^{-1} borate, 0.1% mercaptoethanol; pH 8.8). The homogenate was centrifuged at 10,600× *g* for 15 min at 4 °C, and the supernatant was used for enzyme activity and protein determinations.

The superoxide dismutase (SOD; EC 1.15.1.1) activity based on the measurement of inhibition in the photochemical reduction of nitroblue tetrazolium (NBT) was determined by method of Giannopolitis and Ries [24]. The catalase activity (CAT; EC 1.11.1.6) was analyzed according to Aebi [25]. The glutathione peroxidase (GPX; EC 1.11.1.7) activity was determined as described by Ruley et al. [26].

The concentration of protein was assayed using Bovine serum albumin (BSA) as a standard [27].

Phenylalanine ammonia-lyase activity (PAL; EC 4.3.1.5) was measured according to Mozzetti et al. [28]; 0.2 mL of the extract was added to 2.3 mL of reaction buffer (borate 0.1 M, 10 mM phenylalanine; pH 8.8). The mix was incubated at 40 °C and the reaction stopped, after 60 min, by addition of 0.5 mL of 5 N HCl. The absorbance was read at 290 nm. PAL activity was expressed as the mol of cinnamic acid per kg of tissue produced under the specific conditions.

Malondialdehyde content was determined based on the method used by Heath and Packer [29] and Li, et al. [30]. First, 0.5 g of asparagus was ground to a fine powder with liquid nitrogen and extracted with 1.5 mL of 5% trichloroacetic acid solution. Then, the homogenate was centrifuged at 5000× *g* for 10 min, and the supernatant was diluted to 10 mL. The same volume of 0.67% TBA was mixed with the diluted extract (2 mL). The mixture was incubated for 30 min at 100 °C in water bath and then centrifuged at 5000× *g* for 10 min. The MDA content was expressed as nmol kg^{-1} of FW.

2.6. Sensory Test

A trained panel consisting of five department faculty members conducted the sensory evaluations on the raw packaged spears, after refrigerated storage at 4 ± 1 °C and 90–95% relative humidity. All evaluations took place at the quality and packaging laboratory at 20 °C.

To reduce first-sample and carry-over effects, a randomized distribution was used, and different spears were randomly chosen from each packaging solution, for each assessor.

The sensory analysis test was carried out as reported by Arana et al. [11] and the procedure followed the standard ISO/IEC 17,025 [31].

Firstly, visual references were registered, and spears were evaluated for the following descriptors: 'artichoke-shaped tip', 'firmness', 'color', 'fibrousness', 'growth of mold', 'odors', 'off-odors', and 'overall'. The scale intensity scores were from 0 to 5. The 'artichoke-shaped tip' descriptor was defined as the apical bud with a shape resembling an artichoke, and the evaluation technique consisted of the observation of the tip or bud, noticing the presence of scaly sprouts, and giving 0 to a "close tip", meaning the highest quality; the definition for 'firmness' was the asparagus' resistance to deformation by gravity, and giving 0 was the worst quality. The other descriptors were evaluated using a discontinuous scale from 0 (absence of sensation) to 5 (extremely intense) Supplementary Materials.

In order to award the final quality score of each descriptor considering its relative importance, the quality scores were multiplied by the weighting coefficients, and the 'maximum score' of each descriptor was then the percentage of importance assigned to it with respect to the global quality of the sample [11].

2.7. Statistical Analysis

The data were subjected to a two-way analysis of variance (ANOVA) as a factorial combination of packaging film × storage time. The means were compared with Tukey-test ($p < 0.05$) using CoStat release 6.311 (CoHort Software, Monterey, CA, USA). The sensory data for each attribute were submitted to one-way ANOVA using samples as factors.

3. Results and Discussion

3.1. Effect of Different Packaging and Storage Time on Respiration Rate and Nutritional Quality Parameters

As known, weight loss was significantly higher in the macro-perforated packaging than in the PP C samples: the latter material was able to limit weight loss to 0.67%, compared to 15.8% for PP MA [32]. Statistics underlined the significance of weight loss for packaging and storage as well as their interaction (Table 1) and as confirmed by data, the PP C film is suitable to slow down transpiration, thus limiting the weight loss of the samples. A previous study suggested an oxygen concentration similar to air composition (21%) and a carbon dioxide concentration in the range of 5 to 10% at 5 °C as an adequate atmosphere of storage for asparagus sprouts [33].

Of course, the only modification in headspace composition in our study took place in PP C bags, because PP C had a controlled permeability to gas, while PP MA bags were used to guarantee hygienic condition as well as product containment but allowed non-controlled gas transfer due to macro-perforation. The observed changes in headspace gas composition (Table 1) were determined both by O_2 consumption and by CO_2 production by green asparagus and by the gas permeability through the film.

O_2 values were slightly below the recommended value, ranging around 18% from the 15 days of storage, followed by steady levels; however, CO_2 stabilized around to 3.4% after 15 days. During studies of prolonged shelf life, PP C demonstrated the ability to keep the internal atmosphere close to the suggested value maintaining the good quality of fresh green asparagus.

The performance of chromatic indexes (L*, a*, b*, and ΔE) according to the analysis of variance is reported in Table 1. Packaging influenced only a* ($p \leq 0.01$), b* was influenced by storage time at $p \leq 0.05$, while all other interactions were not statistically significant.

Considering the intricacy of color representation, it is important to remember that the model CIEL*a*b* has a wider range of color, in order to better represent the difference as well as the variation perceived by the human eyes. For this reason, a decrease in one component often corresponds to an increase in another, maintaining stable visual color [34]. During the storage time, from the start of the trial to the end, a small increase in L* values as well as in total color difference was reported. The total color difference (ΔE), as expected, increased during storage despite values in the range 3 to 6 being considered quite distinguishable [34] (Table 1). The negative value of a* indicates a greater tendency towards white color [35].

Ascorbic acid content was highly influenced by storage and packaging conditions (Table 1). It is known that ascorbic acid is one of the most sensitive components and is susceptible to degradation when fresh produce is subject to unsuitable handling and storage conditions [20]. Values found on our samples, and reported in Table 2, at the beginning of the study (0.295 g kg^{-1}) were in line with those reported by other authors, i.e., around 0.190–0.350 g kg^{-1} [19,20].

Samples in PP MA packaging had a quicker degradation of ascorbic acid, with a final content of 0.219 g kg^{-1}, compared to samples packed in PP C which almost maintained the initial concentration, with a loss of 8% after 30 days of storage. Asparagus spears packed into PP C also showed an increase of AsAC during storage (0.349 g kg^{-1} after 22 d) according to a previous study [36]; this path can be explained as a dynamic equilibrium between the starting ascorbic acid value and the production coming from oxidative and senescence processes.

The starting concentration of total polyphenols was 3.0 g kg^{-1} of gallic acid equivalent (Figure 1), higher than values found in other studies [19], reporting values in the range 0.810 to 1.74 g kg^{-1} as a function of ripening, seasonal, and environmental factors and their interactions with agronomic practices as well as the different asparagus cultivars.

Table 1. Effects of "packaging (P)" and "storage time (S)" on qualitative traits, weight loss (%); carbon dioxide (CO_2) and oxygen (O_2) headspace concentrations (%); color parameters (L*, a*, b*, and ΔE); ascorbic acid content (AsAC) (g kg^{-1} FW); total polyphenol content (TPC) (g kg^{-1} FW); chlorophyll-*a* (Chl-*a*); chlorophyll-*b* (Chl-*b*); and total carotenoids (Car) (mg kg^{-1} FW) on green asparagus spears (*Asparagus officinalis* L. 'Vegalim'). Two packaging materials were considered—PP C (polypropylene micro-perforated bags) and PP MA (polypropylene macro-perforated bags).

		Qualitative Trait											
			PP C†										
		WL (%)	CO_2 (%)	O_2 (%)	L*	a*	b*	ΔE	AsAC (g kg^{-1} FW)	TPC (g kg^{-1} FW)	Chl-a (mg kg^{-1} FW)	Chl-b (mg kg^{-1} FW)	Car (mg kg^{-1} FW)
Packaging (P)													
	PP C	0.25 ± 0.06 b	3.62 ± 1.07	17.86 ± 1.61	49.41 ± 3.59	−10.57 ± 0.85 a	26.00 ± 1.55	3.31 ± 1.54	0.29 ± 0.01 a	2.63 ± 0.07 a	200.21 ± 63.6 a	207.76 ± 15.14 a	6.19 ± 0.78 a
	PP MA	5.78 ± 1.14 a	-	-	48.28 ± 3.00	−11.78 ± 0.81 b	26.50 ± 1.39	2.00 ± 0.96	0.26 ± 0.01 b	2.16 ± 0.13 b	183.36 ± 65.34 b	153.45 ± 29.17 b	5.03 ± 1.02 b
Storage time (S)													
	0	0.00 ± 0.00 d	1.00 ± 0.02	21.00 ± 0.10	48.31 ± 1.65	−11.61 ± 0.46	26.95 ± 0.20 a	0.00 ± 0.00	0.29 ± 0.01 a	3.00 ± 0.02 a	667.42 ± 0.98 a	205.38 ± 0.18 ab	20.01 ± 0.00 a
	8	1.43 ± 0.64 cd	6.50 ± 2.60	14.13 ± 4.90	46.89 ± 1.31	−10.95 ± 0.47	25.03 ± 0.53 b	2.87 ± 1.58	0.27 ± 0.00 b	2.54b ± 0.10	88.98 ± 11.15 c	239.20 ± 27.89 a	4.45 ± 1.75 b
	15	3.11 ± 1.37 bc	3.80 ± 0.81	17.91 ± 0.65	47.52 ± 1.20	−10.71 ± 0.19	25.50 ± 0.50 b	2.12 ± 1.16	0.25 ± 0.01 b	2.22 ± 0.17 c	87.28 ± 7.89 c	200.10 ± 8.28 b	1.70 ± 0.39 c
	22	4.58 ± 1.96 ab	3.40 ± 0.43	18.23 ± 0.51	49.96 ± 0.98	−11.44 ± 0.25	26.96 ± 0.17 a	1.90 ± 1.61	0.31 ± 0.02 a	2.23 ± 0.11 c	112.93 ± 9.53 b	103.54 ± 20.11 d	1.88 ± 0.67 c
	30	5.95 ± 2.46 a	3.40 ± 0.45	18.03 ± 0.30	51.56 ± 0.92	−11.19 ± 0.61	26.82 ± 0.18 a	3.71 ± 1.66	0.25 ± 0.01 b	1.98 ± 0.15 d	2.30 ± 1.57 d	154.82 ± 69.58 c	0.00 ± 0.00 d
(P)		***			NS	**	NS	NS	***	***	***	**	***
(S)		***			NS	NS	*	NS	***	***	***	***	***
(P) x (S)		***			NS	NS	NS	NS	***	***	***	***	***

† Headspace carbon dioxide (CO_2) and oxygen (O_2) concentrations were monitored only on polypropylene micro-perforated bags (PP C), because PP MA (polypropylene macro-perforated bags) allowed non-controlled gas transfer. The levels of significance were represented by $p > 0.05$. NS, not significant; * significant at $p < 0.05$; ** significant at $p < 0.01$; *** significant at $p < 0.001$ according to Tukey's test.

Table 2. Effect of the "packaging (P) x storage time (S)" significant interaction on ascorbic acid content (AsAC), chlorophyll-*a* (Chl-*a*), chlorophyll-*b* (Chl-*b*), and carotenoid (Car) in PP C (micro-perforated polypropylene bags) and PP MA (polypropylene macro-perforated bags).

	PP C	PP MA	PP C	PP MA	PP C	PP MA	PP C	PP MA
Storage Time (d)	AsAC ($g\ kg^{-1}FW$)		Chl-*a* ($mg\ kg^{-1}FW$)		Chl-*b* ($mg\ kg^{-1}FW$)		Car ($mg\ kg^{-1}FW$)	
0	0.295 ± 0.28 bA		667 ± 30 aA		205 ± 15 bA		20 ± 0.03 aA	
8	0.266 ± 0.19 bA	0.269 ± 0.15 abA	111 ± 15 bcA	69 ± 18 bB	185 ± 29 bcB	298 ± 37 aA	8.3 ± 0.5 bA	0.6 ± 0.06 cB
15	0.273 ± 0.08 bA	0.231 ± 0.12 bcB	89 ± 26 cA	88 ± 16 bA	194 ± 19 bcA	210 ± 22 bA	0.6 ± 0.00 cB	1.1 ± 0.12 cA
22	0.349 ± 0.05 aA	0.270 ± 0.20 abB	133 ± 15 bA	96 ± 11 bB	150 ± 10 cA	59 ± 9 cB	0.4 ± 0.00 dB	3.4 ± 0.11 bA
30	0.276 ± 0.06 bA	0.219 ± 0.03 cB	5 ± 0.5 d	n.d.	113 ± 27 cA	189 ± 32 bA	n.d.	n.d.

Values are means ± SD of four replicates (n = 4). Different lower-case letters within the column indicate statistically significant differences among storage times according to Tukey's test at the significance level ($p < 0.05$); different capital letters indicate statistically significant differences according to Tukey's test at the significance level ($p < 0.05$). n.d.; not detectable.

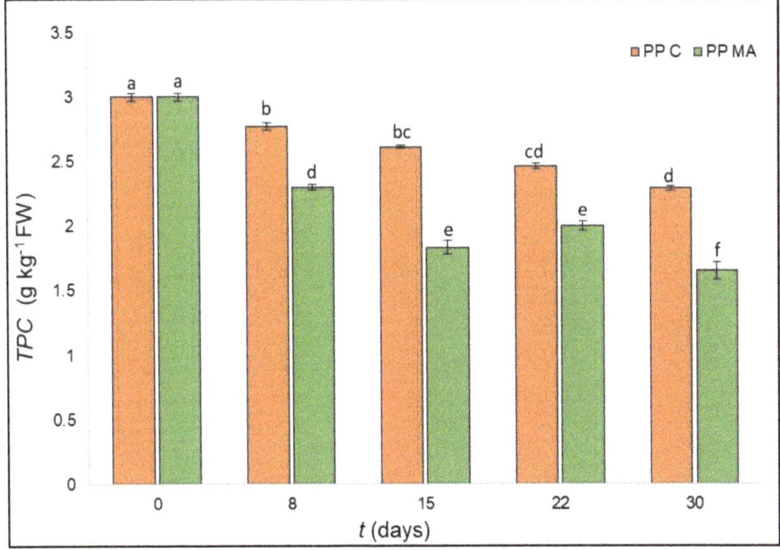

Figure 1. Total phenolic content of green asparagus in PP C and PP MA during 30 days of storage. Data represent the mean of three replicates ± standard deviation. Means with different letters are significantly different ($p < 0.05$) according to Tukey's test.

Total polyphenol content (TPC) was influenced by both storage and packaging conditions tested and by their interaction ($p < 0.001$) (Table 1). TPC decreased for both packaging materials tested: samples packed in PP MA lost 45% of the initial value (1.65 g kg^{-1} after 30 d), while spears packed in PP C limited the decrease to a final value of 2.3 g kg^{-1}, which corresponds to a 24% loss compared to the initial value. It is evident that PP C was able to preserve TPC content during shelf life, considering that the lowest concentration found after 30 days of storage corresponded to the amount found in PP MA after 8 days of storage (Figure 1).

The quality of asparagus spears is differently influenced by the amount of chlorophyll *a* and *b* considering that their ratio inside the chloroplast is 3:1 [37]. The values determined in the fresh product, corresponded exactly to the previous cited ratio, with a starting concentration of 667 mg kg^{-1} in Chl-*a* and 205 mg kg^{-1} for Chl-*b* (Table 2), and they were both influenced by packaging and storage as well as their interaction (Table 1). Moreover, the total chlorophyll concentration also decreased during storage, in both films tested, according to previous studies [20,38].

The Chl-*a* content decreased quickly in the first 8 days for both packaging materials, even reaching a drop of the 90% in samples packed in PP MA (Table 1), probably due the higher gas exchange and faster metabolism, followed by an increase after 15 and 22 days (Table 2), probably due to a decrease in fresh weight caused by a greater water loss of PP MA, which reached undetectable levels after 30 days of storage. In contrast, the Chl-*b* content was slightly more stable during storage, with a more evident drop-off after 22 days in PP MA then in PP C, saving an amount of Chl-*b* in the range of 45 to 80% which was related to water loss (Table 2).

Carotenoids in asparagus spears are not visible, because their color is masked by the green of chlorophylls, but they have a nutritional value if present in low concentrations [38]; their level was influenced by both storage and packaging condition as reported in Table 1. In fresh asparagus spears, the total carotenoid content was 20 mg kg^{-1} (Table 2); after 8 days of storage there was an almost complete (97%) loss in samples packed in PP MA, while samples stored in PP C limited loss to about 58.9% in the same period. Afterwards, at the end of the shelf-life study, the total carotenoid content was not detected.

Considering the possible correlations among color coordinates and the concentration of the pigments analyzed, during the period of storage, they showed a second degree polynomic model. In particular, many of them had the weakest correlation with the color parameters, with an r^2 that did not exceed 0.65. The correlations were slightly higher for the concentrations of Chl-*b* with L* coordinate and of Chl-*a* with a* and ΔE with coefficient of determination of 0.798, 0.828, and 0.983, respectively. Interestingly, the correlation of carotenoid with ΔE had a coefficient of determination equal to 1. However, from all the parameters, ΔE could best be used to estimate the concentration of these pigments. This seems logical considering that this parameter was calculated as it simultaneously utilized the L*, a*, and b* color coordinates, as indicated by its high r^2 values with Chl-*a* and carotenoid.

Texture is one of the main factors in determining eating and cooking quality of fresh asparagus and excessive fiber is a disagreeable quality characteristic. Spear lignification, of the fiber ring and the vascular bundles, in asparagus causes toughening [39]. As reported in Table 3, the analysis of variance showed that packaging and the storage time had a significant ($p < 0.001$) effect for both fiber and lignin parameters.

Table 3. Effects of "packaging" and "storage time" on fiber (%), lignin content (g kg^{-1} FW), phenylalanine ammonia-lyase (PAL) activity (mmol min^{-1} g^{-1}), catalase (CAT), glutathione peroxidase (GPX), and superoxidedismutase (SOD) activity (U mg^{-1} $protein^{-1}$), and malondialdehyde content (MDA) (nmol g^{-1} FW) on green asparagus spears (*Asparagus officinalis* L. 'Vegalim'). Two packaging materials were considered—PP C (polypropylene micro-perforated bags) and PP MA (polypropylene macro-perforated bags).

		Fiber (%)	Lignin (g kg^{-1} FW)	PAL (mmol min^{-1} g^{-1})	CAT (U mg^{-1} $protein^{-1}$)	SOD (U mg^{-1} $protein^{-1}$)	GPX (U mg^{-1} $protein^{-1}$)	MDA (nmol g^{-1} FW)
Packaging (P)								
	PP C	4.98 ± 0.15 b	0.018 ± 7.9 10^{-5} b	0.0096 ± 0.00 b	0.31 ± 0.03 a	25.29 ± 7.7	1034.66 ± 154.9 b	0.74 ± 0.05
	PP MA	5.64 ± 0.36 a	0.021 ± 7.9 10^{-5}a	0.0109 ± 0.00 a	0.24 ± 0.04 b	27.20 ± 7.4	1383.82 ± 203.8 a	0.73 ± 0.02
Storage time (S)								
	0	3.38 ± 0.14 c	0.014 ± 0.02	0.009 ± 0.00 b	0.23 ± 0.05 ab	81.22 ± 5.4 a	708.83 ± 115.6 c	0.77 ± 0.02 ab
	8	4.66 ± 0.18 b	0.019 ± 0.00	0.009 ± 0.00 b	0.34 ± 0.04 a	12.39 ± 1.7 b	859.66 ± 119.0 bc	0.85 ± 0.07 a
	15	6.03 ± 0.16 a	0.020 ± 0.00	0.012 ± 0.00 a	0.028 ± 0.02 ab	10.54 ± 2.0 b	810.98 ± 57.5 bc	0.77 ± 0.03 ab
	22	5.66 ± 0.17 a	0.019 ± 0.00	0.011 ± 0.00 a	0.37 ± 0.01 a	8.16 ± 1.0 b	1308.92 ± 163.2 b	0.65 ± 0.04 b
	30	6.33 ± 0.33 a	0.020 ± 0.00	0.010 ± 0.00 ab	0.15 ± 0.06 b	18.93 ± 0.9 b	2357.81 ± 244.2 a	0.64 ± 0.06 b
(P)		*	***	**	**	NS	**	NS
(S)		***	NS	***	***	***	***	***
(P) x (S)		***	***	***	*	NS	**	**

NS, not significant; *significant at $p < 0.05$; ** significant at $p < 0.01$; *** significant at $p < 0.001$.

As expected, an increase in fiber content was observed for samples in both plastic films tested (Figure 2a). The initial fiber content was 3.38%, a percentage slightly higher compared to values reported in literature, but this is probably a specific characteristic of the 'Vegalim' genotype; among packaging, samples packed in PP C showed an increase of

increase of 53% in PP C compared to PP MA. This parameter declined throughout time, reaching values close to 0 in the PP MA treatment at the end of the trial.

No significant differences in the SOD activity among the packaging and the storage were detected (Table 3). The SOD activity was significantly affected only by packaging. Toivonen and Sweeney, in a study on two genotypes of broccoli, showed that a delay of senescence was correlated to increase of SOD activities [43]. From the results of our study, it can be hypothesized that PP C might delay the senescence of asparagus spears by means of regulating the antioxidant enzyme systems.

Storage time and its interaction with packaging significantly ($p < 0.01$ and $p < 0.001$, respectively) influenced the GPX activity, as evidenced by the ANOVA (Table 3). Our results showed that the GPX activity increased at the end of the trial in both films with an increase of 65% in PP MA and 73% in PP C compared to the initial value (Figure 3c). Asparagus spears treated with PP MA showed, in general, a higher GPX activity than PP C, which might have resulted in efficient conversion of the superoxide anion radical to H_2O_2 then to H_2O [44].

However, SOD and GPX enzyme activities in asparagus are diminished with the lengthening of postharvest time [4], which results in the accumulation of hydrogen peroxide (H_2O_2) in organism [45,46].

The thiobarbituric acid-reactive substance (MDA) is one of final products of polyunsaturated fatty acid oxidation of the cell membrane, resulting as product of membrane lipid peroxidation, which is caused by accumulation of reactive oxygen species and it has been used as a direct indicator of membrane injury. In our study the MDA content remained constant during the storage time in both treatments (Figure 3d). After only 8 days the MDA content of asparagus in PP C was significantly higher (26%) than that stored in PP MA. In several studies of fruit and vegetable shelf life, a continuous increase in MDA content was also observed; for post-harvest broccoli it was explained [47,48] that senescence is related to lipid peroxidation (MDA content) resulting to disintegration of the cell membrane.

3.3. Sensorial Test

Table 4 shows the mean scores of the significant descriptive sensory analysis attributes of 'Vegalim' green asparagus in the two different packaging materials. As can be observed, no significant difference between treatments was detected in the first 15 days of storage for any of the eight quality attributes considered. After 22 days of storage the descriptors which recorded the most significant differences were turgidity, firmness, color, and total acceptability. These scored better values for PP C than for PP MA, with artichoke-shaped tip and odors being identified as different. It is noteworthy that the statistically significant changes highlighted by instrumental color parameters did not correspond to perceived color differences until 22 days of storage. This could be due to samples' variability and to the small, though significant, differences recorded by the colorimeter. At the end of the shelf-life study, the main differences were, again, in firmness, color, turgidity, and total acceptability, the same four parameters which showed significant differences in the previous sampling.

In this last sampling (30 days) the judges' assessments were similar to the previous ones for PP MA, while the values of samples in PP C were noticeably worse. The worst attributes, mold growth and off-odors, were not detected by judges, proving the genotype 'Vegalim' suitable for prolonged refrigerated storage in both packaging materials tested.

Overall, the comparison between the two packaging materials showed that a film with controlled gas permeation rather than a macro-perforated film is more suitable for quality maintenance of fresh asparagus spears by reducing water vapor transmission, thus creating a saturated moisture atmosphere which minimizes transpiration and consequent withering. Similarly, the PP C film was able to generate a suitable headspace, close to values suggested in literature, which resulted from the combination of the vegetable metabolism (O_2 uptake and CO_2 emission) and film permeability (CO_2 flowing outwards and O_2 inwards), following partial pressure gradients. The resulting passive atmosphere has positive effects on the metabolism of the vegetable, as was confirmed by the better retention of bioactive components and highest sensory scores.

Table 4. Mean scores of the significant sensory attributes of asparagus in different packaging materials—PP C (micro-perforated coextruded PP film) and PP MA (macro-perforated PP film) during 30 days of storage.

Storage Time (d)	Attribute	Packaging	
		PP C	*PP MA*
0		No significant differences	
8		No significant differences	
15		No significant differences	
22	Artichoke-shaped tip	1.8 b	4.2 a
	Firmness	5.0 a	2.4 b
	Color	4.0 a	1.8 b
	Turgidity	4.0 a	2.0 b
	Odors	3.6 a	1.2 b
	Total acceptability	3.6 a	1.2 b
30	Firmness	4.2 a	2.4 b
	Color	4.0 a	2.6 b
	Turgidity	4.0 a	2.4 b
	Total acceptability	4.0 a	2.2 b

* Values marked with different letters in the same row are significantly different ($p < 0.05$) according to the Tukey's test.

4. Conclusions

Despite many previous papers studying different postharvest treatments for maintaining the quality of asparagus spears, and focusing on changes in packaging materials and storage conditions such as temperature or relative humidity, or modifying the atmosphere, changing oxygen, carbon dioxide, and nitrogen percentage, few of them have considered a prolonged shelf life as we did in our study. A longer shelf life has a double-positive effect both for limiting food waste and extending the economic value for the presence on the market. Since, nowadays, asparagus spears are often distributed unpacked, in bunches surrounded by paperboard or plastic bands, it is relevant to assess the potential of packaging materials commonly used for fruits and vegetables (polyethylene, low density polyethylene, and polypropylene) for shelf-life extension. PP C film was confirmed as the best choice between our tested plastic films. PP C package could effectively limit water loss and maintain the fresh appearance of fresh asparagus spears during storage, thus allowing an extension in postharvest life. It controlled the weight loss and kept, naturally, the internal atmosphere close to suggested values for fresh green asparagus according to literature data. Also, from a chemical point of view, PP C had the best performance. Asparagus spears packed in PP C could slow down carotenoid loss, reduce PAL, GPX and MDA activity after 30 days of refrigerated storage and improve sensorial qualities compared with PP MA. In PP MA, the observed AsAC, TPC, fiber, and lignin concentrations are partially determined by a significant level of dehydration, as proved by weight loss, which was almost negligible in PP C. In other words, the observed difference would be even greater if we considered those parameters on dry weight basis. Mold growth and off-odors were not detected, proving 'Vegalim' asparagus as a suitable variety for prolonged refrigerated storage. Providing this genotype with a suitable packaging such as PP C would greatly improve its potential for distribution by retarding senescence and maintaining higher quality parameters.

Research on packaging materials is nowadays highly focused on biodegradable materials, especially for fresh products packaging. However, the use of conventional plastic materials still offers higher convenience, especially when mono-material films are concerned. Indeed, PP films can be effectively recovered and recycled in the current waste management systems, while biodegradable films, to date, represent an issue for the current composting facilities. PP C film represents a viable strategy for shelf life extension of green asparagus and it can thus represent a reference material for future evaluation of biodegradable/compostable materials once cost, performance, and waste management aspects are improved.

Supplementary Materials: The following are available online at https://www.mdpi.com/2304-8158/10/2/478/s1.

Author Contributions: Conceptualization: G.M.; D.R., V.R. and F.L; formal analysis: S.T., V.R. and F.L.; writing—original draft: S.T. and V.R.; writing—review and editing: V.R., S.T., D.R., F.L. and G.M. All authors have read and agreed to the published version of the manuscript.

Funding: The research was supported by the Project "Sostanze naturali e colture starter per la qualità e la sicurezza microbiologica degli alimenti" funded by the University of Catania, Italy.

Institutional Review Board Statement: Not applicable.

Informed Consent Statement: Not applicable.

Data Availability Statement: Main data are contained within the article or supplementary material; further data presented in this study are available on request from the corresponding author.

Acknowledgments: To Asparago Sovrano, asparagus producer consortium (C.da Imbaccari Sovrano, Piazza Armerina, Enna, Italy) for the spear supply.

Conflicts of Interest: The authors declare no conflict of interest.

References

1. Fuentes-Alventosa, J.M.; Rodríguez-Gutiérrez, G.; Jaramillo-Carmona, S.; Espejo-Calvo, J.A.; Rodríguez-Arcos, R.; Fernández-Bolaños, J.; Guillén-Bejarano, R.; Jiménez-Araujo, A. Effect of extraction method on chemical composition and functional characteristics of high dietary fibre powders obtained from asparagus by-products. *Food Chem.* **2009**, *113*, 665–671. [CrossRef]
2. Garcia-Herrera, P.; Sànchez-Mata, M.C.; Càmara, M.; Tardìo, J.; Olmedilla-Alonso, B. Carotenoid content of wild edible young shoots traditionally consumed in Spain (*Asparagus acutifolius* L., *Humulus lupulus* L., *Bryonia dioica* Jacq. and *Tamus communis* L.). *J. Sci. Food Agric.* **2013**, *93*, 1692–1698. [CrossRef] [PubMed]
3. Villanueva, M.J.; Tenorio, M.D.; Sagardoy, M.; Redondo, A.; Saco, M.D. Physical, chemical, histological and microbiological changes in fresh green asparagus (*Asparagus officinalis* L.) stored in modified atmosphere packaging. *Food Chem.* **2005**, *91*, 609–619. [CrossRef]
4. An, J.; Zhang, M.; Lu, Q. Changes in some quality indexes in fresh-cut green asparagus pretreated with aqueous ozone and subsequent modified atmosphere packaging. *J. Food Eng.* **2007**, *78*, 340–344. [CrossRef]
5. Chang, D.N. *Asparagus. Postharvest Physiology of Vegetables*; Weichmann, J., Ed.; Marcel Dekker: New York, NY, USA, 1987; pp. 523–525.
6. Kader, A.A. Postharvest biology and technology: An overview. In *Postharvest Technology of Horticultural Products*, 2nd ed.; Kader, A.A., Ed.; UC Davis: Davis, CA, USA, 1992; pp. 15–20.
7. Haihong, Y.; Zeng, J.; Kumano, T.; Fujii, M.; Araki, H. New asparagus cultivars suitable for outdoor cultivation in Hokkaido, Japan. *Acta Hortic.* **2018**, *1223*, 25–32. [CrossRef]
8. Bhowmik, P.K.; Matsui, T.; Kawada, K.; Suzuki, H. Seasonal changes of asparagus spears in relation to enzyme activities and carbohydrate content. *Sci. Hortic.* **2001**, *88*, 1–9. [CrossRef]
9. Bhowmik, P.K.; Matsui, T.; Ikeuchi, T.; Suzuki, H. Changes in storage quality and shelf life of green asparagus over an extended harvest season. *Postharvest Biol. Technol.* **2002**, *26*, 323–328. [CrossRef]
10. Lipton, W.J. Postharvest biology of fresh asparagus. *Hortic. Rev.* **1990**, *12*, 69–155. [CrossRef]
11. Arana, I.; Ibañez, F.C.; Torre, P. Monitoring the sensory quality of canned white asparagus through cluster analysis. *J. Sci. Food Agric.* **2015**, *96*, 2391–2399. [CrossRef]
12. Wang, J.; Fan, L. Effect of ultrasound treatment on microbial inhibition and quality maintenance of green asparagus during cold storage. *Ultrason. Sonochem.* **2019**, *58*, 104631. [CrossRef]
13. Mittler, R. Oxidative stress, antioxidants and stress tolerance. *Trends Plant Sci.* **2002**, *7*, 405–410. [CrossRef]
14. Han, D.H.; Lee, M.J.; Kim, J.H. Antioxidant and apoptosis-inducing activities of ellagic acid. *Anticancer. Res.* **2006**, *26*, 3601–3606. [PubMed]
15. Muratore, G.; Restuccia, C.; Licciardello, F.; Lombardo, S.; Pandino, G.; Mauromicale, G. Effect of packaging film and anti-browning solution on quality maintenance of minimally processed globe artichoke heads. *Innov. Food Sci. Emerg. Technol.* **2015**, *31*, 97–104. [CrossRef]
16. Licciardello, F.; Muratore, G.; Spagna, G.; Branca, F.; Ragusa, L.; Caggia, C.; Randazzo, C.; Restuccia, C. Evaluation of some quality parameters of minimally processed violet and white-type cauliflowers. *Acta Hortic.* **2013**, *1005*, 315–321.

17. Licciardello, F.; Lombardo, S., Rizzo, V., Pitino, I.; Pandino, G.; Strano, M.G.; Muratore, G.; Restuccia, C.; Mauromicale, G. Integrated agronomical and technological approach for the quality maintenance of ready-to-fry potato sticks during refrigerated storage. *Postharvest Biol. Technol.* **2018**, *136*, 23–30. [CrossRef]
18. Association of Official Analytical Chemists. Vitamin C in juices and vitamin preparations. Official Method 967.21. In *AOAC Official Methods of Analysis*, 18th ed.; Association of Official Analytical Chemists: Gaithersburg, MD, USA, 2005; pp. 45–114.
19. Shou, S.; Lu, G.; Huang, X. Seasonal variations in nutritional components of green asparagus using the mother fern cultivation. *Sci. Hortic.* **2007**, *112*, 251–257. [CrossRef]
20. Li, T.; Zhang, M. Effects of modified atmosphere package (MAP) with a silicon gum film window on the quality of stored green asparagus (*Asparagus officinalis* L.) spears. *LWT Food Sci. Technol.* **2015**, *60*, 1046–1053. [CrossRef]
21. Lichtenthaler, H.K. Chlorophylls and carotenoids, the pigments of photosynthetic biomembranes. In *Methods in Enzymology*; Douce, R., Packer, L., Eds.; Academic Press Inc.: New York, NY, USA, 1987; pp. 350–382.
22. Sothornvit, R.; Kiatchanapaibul, P. Quality and shelf-life of washed fresh-cut asparagus in modified atmosphere packaging. *LWT Food Sci. Technol.* **2009**, *42*, 1484–1490. [CrossRef]
23. Techavuthiporn, C.; Boonyaritthongchai, P. Effect of prestorage short-term Anoxia treatment and modified atmosphere packaging on the physical and chemical changes of green asparagus. *Postharvest Biol. Technol.* **2016**, *117*, 64–70. [CrossRef]
24. Giannopolitis, C.N.; Ries, S.K. Superoxide dismutases: I. Occurrence in higher plants. *Plant Physiol.* **1977**, *59*, 309–314. [CrossRef]
25. Aebi, H. Catalase in Vitro. *Methods Enzymol.* **1984**, *105*, 121–126. [CrossRef]
26. Ruley, A.T.; Sharma, N.C.; Sahi, S.V. Antioxidant defense in a lead accumulating plant, *Sesbania drummondii. Plant Physiol. Biochem.* **2004**, *42*, 899–906. [CrossRef]
27. Bradford, M.M. Rapid and Sensitive Method for the Quantitation of Microgram Quantities of Protein Utilizing the Principle of Protein-Dye Binding. *Anal. Biochem.* **1976**, *72*, 248–254. [CrossRef]
28. Mozzetti, C.; Ferraris, L.; Tamietti, G.; Matta, A. Variation in enzyme activities in leaves and cell suspensions as markers of incompatibility in different *Phytophthora*-pepper interactions. *Physiol. Mol. Plant Path.* **1995**, *46*, 95–107. [CrossRef]
29. Heath, R.L.; Packer, L. Photoperoxidation in isolated chloroplasts. I. Kinetics and stoichiometry of fatty acid peroxidation. *Arch. Biochem. Biophys.* **1986**, *125*, 189–198. [CrossRef]
30. Li, G.; Wan, S.; Zhou, J.; Yang, Z.; Qin, P. Leaf chlorophyll fluorescence, hyperspectral reflectance, pigments content, malondialdehyde and proline accumulation responses of Castor Bean (*Ricinus communis* L.) seedlings to salt stress levels. *Ind. Crop Prod.* **2010**, *31*, 13–19. [CrossRef]
31. International Organization for Standardization (ISO). *General Requirements for the Competence of Testing and Calibration Laboratories*; ISO/IEC 17025:2005; ISO: Geneva, Switzerland, 2005. Available online: https://www.iso.org/standard/66912.html (accessed on 3 August 2003).
32. Rizzo, V.; Chiella, G.; Licciardello, F.; Toscano, S.; Romano, D.; Muratore, G. Shelf life assessment of green asparagus packaged in polypropylene macro-perforated and micro-perforated. "Shelf Life International Meeting 2017 (SLIM 2017)", Bangkok (Thailand), 1–3 November 2017. *Ital. J. Food Sci.* **2017**, *22*, 115–119. [CrossRef]
33. Kader, A.A. Ethylene-induced senescence and physiological disorders in harvested horticultural crops. *HortScience* **1985**, *20*, 54–57.
34. Riva, M. Il Colore Degli Alimenti e la sua Misurazione. diSTAM Info Sheet. 2003. Available online: http://users.unimi.it/~{}distam/info/colore.pdf (accessed on 3 August 2003).
35. Siomos, A.S.; Sfakiotakis, E.M.; Dogras, C.C. Modified atmosphere packaging of white asparagus spears: Composition, color and textural quality responses to temperature and light. *Sci. Hortic.* **2000**, *84*, 1–13. [CrossRef]
36. Reyes, L.F.; Villareal, J.E.; Cisneros-Zevallos, L. The increase in antioxidant capacity after wounding depends on the type of fruit or vegetable tissue. *Food Chem.* **2007**, *101*, 1245–1262. [CrossRef]
37. Gaur, S.; Shivhare, U.; Ahmed, J. Degradation of chlorophyll during processing of green vegetables: A review. *Stewart Postharvest Rev.* **2006**, *5*, 1–8. [CrossRef]
38. Mastropasqua, L.; Tanzarella, P.; Paciolla, C. Effect of postharvest light spectra on quality and health- related parameters in green *Asparagus officinalis* L. *Postharvest Biol. Technol.* **2006**, *112*, 143–151. [CrossRef]
39. Siomos, A.S.; Gerasopoulos, D.; Tsouvaltzis, P.; Koukounaras, A. Effects of heat treatment on atmospheric composition and color of peeled white asparagus in modified atmosphere packaging. *Innov. Food Sci. Emerg. Technol.* **2010**, *11*, 118–122. [CrossRef]
40. Toscano, S.; Ferrante, A.; Leonardi, C.; Romano, D. PAL activities in asparagus spears during storage after ammonium sulfate treatments. *Postharvest Biol. Technol.* **2018**, *140*, 34–41. [CrossRef]
41. Wu, L.; Yang, H. Combined application of carboxymethyl chitosan coating and brassinolide maintains the postharvest quality and shelf life of green asparagus. *J. Food Proc. Preserv.* **2016**, *40*, 154–165. [CrossRef]
42. Apel, K.; Hirt, H. Reactive oxygen species: Metabolism, oxidative stress, and signal transduction. *Annu. Rev. Plant Biol.* **2004**, *55*, 373–399. [CrossRef]
43. Toivonen, P.M.A.; Sweeney, M. Differences in chlorophyll loss at 13 C for two broccoli (*Brassica oleracea* L.) cultivars associated with antioxidant enzyme activities. *J. Agric. Food Chem.* **1998**, *46*, 20–24. [CrossRef] [PubMed]
44. Foye, C.H.; Noctor, G. Oxidant and antioxidant signalling in plants: A re-evaluation of the concept of oxidative stress in a physiological context. *Plant Cell Environ.* **2005**, *28*, 1056–1071. [CrossRef]

45. Dhindsa, R.S.; Plumb-Dhindsa, P.; Thorpe, T.A. Leaf senescence: Correlated with increased levels of membrane permeability and lipid peroxidation, and decreased levels of superoxide dismutase and catalase. *J. Exp. Bot.* **1981**, *32*, 93–101. [CrossRef]
46. Droillard, M.J.; Paulin, A.; Massot, J.C. Free radical production, catalase and superoxide dismutase activities and membrane integrity during senescence of petals of cut carnations (*Dianthus caryophyllus*). *Physiol. Plant.* **1987**, *71*, 197–202. [CrossRef]
47. Zhuang, H.; Hildebrand, D.F.; Barth, M.M. Temperature influenced lipid peroxidation and deterioration in broccoli buds during postharvest storage. *Postharvest Biol. Technol.* **1997**, *10*, 49–58. [CrossRef]
48. Zhuang, H.; Hildebrand, D.F.; Barth, M.M. Senescence of broccoli buds is related to changes in lipid peroxidation. *J. Agric. Food Chem.* **1995**, *43*, 2585–2591. [CrossRef]

MDPI

Article

Improved Shelf-Life and Consumer Acceptance of Fresh-Cut and Fried Potato Strips by an Edible Coating of Garden Cress Seed Mucilage

Marwa R. Ali [1,*], Aditya Parmar [2], Gniewko Niedbała [3], Tomasz Wojciechowski [3], Ahmed Abou El-Yazied [4], Hany G. Abd El-Gawad [4], Nihal E. Nahhas [5], Mohamed F. M. Ibrahim [6] and Mohamed M. El-Mogy [7]

1 Food Science Department, Faculty of Agriculture, Cairo University, Giza 12613, Egypt
2 Natural Resources Institute, University of Greenwich, Central Avenue, Chatham Maritime, Kent ME4 4TB, UK; A.Parmar@gre.ac.uk
3 Department of Biosystems Engineering, Faculty of Environmental and Mechanical Engineering, Poznań University of Life Sciences, Wojska Polskiego 50, 60-627 Poznań, Poland; gniewko.niedbala@up.poznan.pl (G.N.); tomasz.wojciechowski@up.poznan.pl (T.W.)
4 Department of Horticulture, Faculty of Agriculture, Ain Shams University, Cairo 11566, Egypt; ahmed_abdelhafez2@agr.asu.edu.eg (A.A.E.-Y.); hany_gamal2005@agr.asu.edu.eg (H.G.A.E.-G.)
5 Department of Botany and Microbiology, Faculty of Science, Alexandria University, Alexandria 21515, Egypt; nihal.elnahhas@alexu.edu.eg
6 Department of Agricultural Botany, Faculty of Agriculture, Ain Shams University, Cairo 11566, Egypt; ibrahim_mfm@agr.asu.edu.eg
7 Vegetable Crops Department, Faculty of Agriculture, Cairo University, Giza 12613, Egypt; elmogy@agr.cu.edu.eg
* Correspondence: marwa3mrf@agr.cu.edu.eg; Tel.: +20-100-465-9915

Citation: Ali, M.R.; Parmar, A.; Niedbała, G.; Wojciechowski, T.; Abou El-Yazied, A.; El-Gawad, H.G.A.; Nahhas, N.E.; Ibrahim, M.F.M.; El-Mogy, M.M. Improved Shelf-Life and Consumer Acceptance of Fresh-Cut and Fried Potato Strips by an Edible Coating of Garden Cress Seed Mucilage. *Foods* **2021**, *10*, 1536. https://doi.org/10.3390/foods10071536

Academic Editors: Valeria Rizzo and Muratore Giuseppe

Received: 8 June 2021
Accepted: 28 June 2021
Published: 2 July 2021

Abstract: Coatings that reduce the fat content of fried food are an alternate option to reach both health concerns and consumer demand. Mucilage of garden cress (*Lepidium sativum*) seed extract (MSE) was modified into an edible coating with or without ascorbic acid (AA) to coat fresh-cut potato strips during cold storage (5 °C and 95% RH for 12 days) and subsequent frying. Physical attributes such as color, weight loss, and texture of potato strips coated with MSE solutions with or without AA showed that coatings efficiently delayed browning, reduced weight loss, and maintained the texture during cold storage. Moreover, MSE with AA provided the most favorable results in terms of reduction in oil uptake. In addition, the total microbial count was lower for MSE-coated samples when compared to the control during the cold storage. MSE coating also performed well on sensory attributes, showing no off flavors or color changes. As a result, the edible coating of garden cress mucilage could be a promising application for extending shelf-life and reducing the oil uptake of fresh-cut potato strips.

Keywords: *Lepidium sativum*; potato; browning index; oil uptake; antioxidant activity

1. Introduction

Potato (*Solanum tuberosum* L.) is one of the most popular crops globally. It is the fourth largest crop in production, after rice, wheat, and maize [1]. In 2019, and according to FAO statistics, the world production of potatoes was around 400 million tons from 19.25 million ha. Apart from its importance as a significant carbohydrate supplier, potato is an excellent source of several essential minerals, vitamins, dietary fiber, and antioxidants [2]. The demand for fresh-cut, minimal processed, and ready-to-eat fruits and vegetables, including potatoes, is increasing, particularly in urban areas due to the convenient nature of these products. During processing, the unit operations (peeling, cutting, and slicing) cause severe damage to living tissues, which increases the respiration rate, enzymatic browning, microbial spoilage, and water loss, resulting in quality losses and reduced shelf-life [3,4].

Enzymatic browning is one of the most prevalent causes of deterioration and loss of quality in processed potato products. Enzymatic browning negatively affects the sensory quality and nutritional components of potatoes [5]; it is triggered when phenolic compounds are leaked from damaged tissues and oxidized by polyphenol oxidase on the surface of fresh-cut potatoes [6]. Therefore, inhibiting enzymatic browning by various pre-treatments has been an important topic of research. Various pre-treatments have been studied for preventing enzymatic browning in potatoes, for example, dipping in hot water (55 °C) for 10 min [7], ultrasonic treatment [8], exogenous γ-aminobutyric acid (20 g L^{-1}) application [9], and the application of 3-mercapto-2-butanol under the concentration of 25 $\mu L\ L^{-1}$ at 5 °C [10]. Moreover, the application of antioxidants (oxalic and ascorbic acid) is well known to control enzymatic browning [11,12]. However, some of these anti-browning agents and treatments are generally not practical in industrial application due to their low customer acceptance, high cost, and negative impact on physicochemical properties.

Several dishes are prepared from potato in almost different cultural and geographical regions. However, one of the most common preparations is deep-fried potato chips and fries [13]. Consumer preferences are changing towards convenient food, as they look for healthier choices with low sugar and fat content [14]. Reducing oil uptake for fried foods by pre-drying before frying, frying under high temperature for a short time, and using edible coatings has been a subject of interest in the previous research [15]. The edible coating to reduce oil update in fried foods can provide a practical alternative for commercial applications. However, the research and development in the edible coating to reducing oil uptake are still in their infancy. This article contributes to developing edible coatings as a mainstream application to reduce the oil content of fried foods such as potato chips and fries.

Lepidium sativum is known as garden cress or pepper cress and belongs to the family *Bassicaceae* [16]. The various parts of the garden cress plant, such as seeds, leaves, or aerials, contain different phytochemicals (flavonoids, glycosides, alkaloid, and polyketides) and vitamins, minerals, proteins, fats, and carbohydrates. The principal constituents of garden cress are linolenic acid (33%) and oleic acid as the most essential fatty acids (23%), the principal sterol is β sitosterol (50%), and tocopherol (1.5–1.9 g kg^{-1}) with c-tocopherol [17]. Due to these phytochemicals, garden cress is known for its various therapeutic and herbal effects [18]. Garden cress gum, categorized as a kind of water-soluble hydrocolloid, might be a potential raw material for preparing edible films and coatings. This is due to its physical, mechanical, optical, and barrier characteristics, which are essential in food packaging applications [19]. The biodegradable, edible nature of the garden cress seed gum coating and the seeds' characterization (physical, microstructural, mechanical thermal characteristics) have been studied before [20]. A significant quality (microbiological, chemical, and sensorial) and shelf-life (cold storage) improvement was observed in shrimp samples coated with garden cress seed gum that contains 10% carvacrol [21]. The major advantages of edible films and coatings are biocompatibility, eco-friendliness (reducing 66% of total packaging wastes), low cost, and excellent barrier properties for gases, lipid, and aroma. Moreover, edible coatings can also be used as a carrier for foods additives and natural bioactive compounds such as vitamins, antioxidants, and antimicrobial compounds [3,22].

This study aimed to evaluate the effect of edible coating based on garden cress (*Lepidium sativum*) seed mucilage (gum) with or without ascorbic acid on the shelf life, microbial load, physical properties, oil uptake, and consumer acceptance of fresh-cut potato strips during cold storage at 5 °C for 12 days, which were subsequently fried into potato fries.

2. Materials and Methods

2.1. Materials

Lepidium sativum (Golden grass) seeds were purchased from the local market in Giza, Egypt. Methyl alcohol, gallic acid, sodium carbonate, 2,2-Diphenyl-1-picrylhydrazyl "DPPH" and Folin and Ciocalteu's phenol reagent were purchased from Sigma-Aldrich

(USA). Potato dextrose agar, plate count agar, and ethylene methylene blue agar were purchased from Oxoid Ltd. (Basingstoke, UK).

2.2. Water Extraction of Lepidium Sativum Seeds Mucilage

The mucilage was extracted according to Karazhiyan et al. [23] with some modifications. First, 50 g of seeds were soaked in 500 mL of distilled water for 12 h and followed by blending (Blender, Moulinex 400 W, Model: LM2420, French) for 15 min at 4000 RPM. Next, blended seeds were filtered under vacuum using the Buchner flask. The pH (pH meter: Jenway, model 3305, Dunmow, Essex, UK) of mucilage seed extract (MSE) was 4.7. For preparing the coating solution, 0.5% glycerol was added to seed mucilage extract and then pasteurized at 90 °C for 1 min.

2.3. Total Phenolic Compounds of Seedcake and MSE

The phenolic content in the fruit juices was estimated by the Folin–Ciocalteu method as described by Awad et al. [4] with some modifications. Five grams of homogenized *L. sativum* seedcake or 5 mL mucilage seed extract was extracted with 50 mL of methanol 80% in a conical flask with a shaker at 1000 rpm for 1 h at room temperature. The extract was then filtered with filter paper No 1; 0.5 mL of the extract was mixed with 2.5 mL of Folin–Ciocalteu reagent (1:10 with water) and, after 3 min, 2 mL of sodium carbonate (7.5%) was added. The absorbance was measured at 765 nm after 1 h of incubation in the dark at room temperature by a spectrophotometer (Unico UV-2000, UNICO company, Fairfield, NJ, USA). TPC was expressed as the gallic acid (GAE g kg^{-1}) dry weight of seeds or 1 L of MSE.

2.4. Antioxidant Activity % of MSE

The antioxidant activity of MSE was determined according to Ali and El Said [24]. First, 1 mL of the MSE was added to 3 mL of methanol and 1 mL of 2,2-diphenyl-1-picrylhydrazyl (DPPH) (0.024 g DPPH in 100 mL^{-1} of methanol). The mixture was incubated in the dark at room temperature for 30 min, followed by absorbance measurements at 517 nm. The antioxidant activity was expressed as % of activity (Equation (1)).

$$Activity\ (\%) = \left[\frac{A\ control - A\ sample}{A\ control}\right] \times 100 \quad (1)$$

A control: the absorbance of the control.
A sample: the absorbance of the sample.

2.5. Preparation of Coated Fresh-Cut Potato Strips

The potatoes were hand-peeled, cut into strips (1.0 × 1.0 × 8 cm) with a manual French fry cutter and washed with cold water to remove excess starch. The potato strips were then drained to reduce the water content on the surface. The samples were divided into 5 treatments as follow:

1. Control: the strips were dipped in distilled water for 5 min.
2. Ascorbic acid (AA): the strips were dipped in AA solution with a concentration of 500 ppm for 5 min.
3. Mucilage seed extract (MSE): the strips were dipped in MSE-coating solution for 5 min.
4. Mucilage seed extract with ascorbic acid (MSE + AA): the strips were dipped in MSE + AA (250 ppm) coating solution for 5 min.
5. Blanching: the strips were blanched for 2 min at 97 °C.

The strips were dipped at room temperature using gentle magnetic agitation (the ratio of grams of potato tissue to milliliters of the solution was 1:4). The strips were allowed to dry at ambient conditions and then placed in polypropylene trays covered with polypropylene pouches (each pouch contained 200 g potato strips ("approx. 20 potato strips"), which were closed thermally. The samples were stored at 5 °C and 95% RH for 12 days. The strips potato samples were taken at 4 d intervals for physical, chemical, and microbiological analysis. Specifically, we used 3 pouches as replicates for each treatment, so totally we prepared 9 pouches for each treatment.

2.6. Frying of Potato Strips

All samples were deep fried according to Reis et al. [25] in a deep frier (Orbit Pan 2000 Watts, Model: 2724297419595, Turkey) with a capacity of three liters of refined canola oil (1:6 strips: oil ratio) at 170 °C for 4 min. The oil was changed every time each batch was fried. Finally, the fried strips were allowed to cool and analyzed for oil uptake at room temperature.

2.7. Determination of Weight Loss, Texture, and Browning Index

Potato strips were weighed immediately after drying and at 0, 4, 8, and 12 days. The results are shown as the percentage weight loss compared to the initial fresh weight [26]. Ten potato strips from each treatment were used to determine texture using a digital penetrometer (PCE-PTR 200, PCE Americas Inc., Jupiter, FL, USA), and values are presented as Newtons (N). The browning index (BI) was calculated according to El-Mogy et al. [3] using Equation (2).

$$BI = \frac{[100\,(x - 0.31)]}{0.17} \tag{2}$$

where:

$$x = \frac{(a^* + 1.75L^*)}{(5.645L^* + a^* - 0.3012b^*)}$$

A Minolta colorimeter (Model CR-400 Chroma Meter, Konica Minolta, INC, Tokyo, Japan) was used to measure a^* (color change from red to green), b^* (color change from yellow to blue), and L^* (lightness) values. A standard white calibration plate was used to calibrate the colorimeter [27].

2.8. Total Phenolic Compounds (TPC) of Potato Strips

The total phenolic compounds were determined according to the Folin–Ciocalteu spectrophotometric method described by Awad et al. [3] for potato strips, *L. sativum* seedcake and mucilage seed extract (MSE). Five grams of homogenized potato strips (interval during storage time) were extracted with 50 mL of 80% methanol in a conical flask with a shaker at 1000 rpm for 1 h at room temperature. The extract was then filtered with filter paper No 1., and 0.5 mL of potato extract was mixed with 2.5 mL of Folin–Ciocalteu reagent (1:10 with water), and after 3 min, 2 mL of sodium carbonate (7.5%) was added. The absorbance was measured at 765 nm after 1 h of incubation in the dark at room temperature. TPC was expressed as the gallic acid (GAE mg 100 g^{-1}) fresh weight of potato strips.

2.9. Oil Uptake (OU)%

Fat content (FC)% of potato French fries were determined according to a Bligh and Dyer [28] extraction. The OU% in the coated potato French fries relative to the uncoated ones was expressed as in Equation (3):

$$OU\ (\%) = \frac{FC_{coated}}{FC_{uncoated}} \times 100 \tag{3}$$

In addition, the reduction of oil uptake was calculated by Equation (4).

$$Oil\ uptake\ reduction\ (\%) = \frac{FC_{coated} - FC_{uncoated}}{FC_{coated}} \times 100 \tag{4}$$

where FC = fat content (%).

2.10. Microbiological Analysis

Ten grams of potato strip samples were crushed and diluted (1:10 *w/v*) in 0.1% buffered peptone water, homogenized by hand massaging for 3 min, and serially diluted with buffered peptone water. The total count was determined using the standard plate count method described by Shehata et al. [29] using plate count agar media and incubating plates at 30 ± 1 °C for 48 h. Mold and yeast counts were performed in potato dextrose agar by incubation at 25–28 °C for 5–7 days. *Escherichia coli* was determined according to ISO 16654 [30]. Microbial counts were expressed as log CFU g^{-1} of tissue.

2.11. Statistical Analysis

The experiments were performed using a completely randomized design. R version 4.0.2 Statistical Package (Vienna, Austria) was used for data analysis. A two-way ANOVA (analysis of variance) for measures was conducted with holm-corrected LSD tests for CLD letters. In addition, a non-metric multidimensional scaling (NMDS) for sensory parameters was performed having the first two dimensions of meta-MDS (comm = ff, k = 3) (multidimensional global scaling using mono-MDS).

2.12. Sensory Analysis

Sensory characteristics including color, taste, odor, texture, and overall acceptability were evaluated for fresh cut and fried potato strips once immediately after treatment by 50 untrained panelists of the Food Science Department (35 females and 15 males, aged 22 to 45 y). A 9-point Hedonic scale (0–2 = dislike extremely, 3–4 = dislike slightly, 5 = fair, 6–8 = like moderately, and 9 = excellent) was utilized for this purpose [31]. The potato strip samples were served at room temperature on two plates: fresh cut and fried potato strips.

3. Results and Discussion

3.1. Weight Loss, Texture and Browning Index

All treatments lost significantly less weight than the controls, starting from 4 days of storage until the end of the storage period (Figure 1). The lowest weight loss values were obtained from MSE + AA treatment. Similar effects of MSE + AA were shown previously on reducing weight loss in vegetables and fruits such as artichoke heads [3] and plums [32]. The beneficial effects of coating for decreasing weight loss might be due to adjusting the internal atmosphere and reduced respiration rates [3,29]. In addition, ascorbic acid contribution to the reduction in weight loss is possible due to the detoxification of active oxygen species [33].

As shown in Figure 2A, the texture of the control decreased with increasing storage time. The samples lost about 75% of their firmness after 12 days from the start of the storage.

Blanching led to lower firmness values than all treatments and the control starting from zero time until the end of storage. Moreover, the firmness of potato strips treated with MSE coating alone or with AA also decreased, but the decrease was significantly smaller than in other treatments. The firmness of fresh-cut fruits and vegetables is one of the most critical factors affecting the quality and shelf life [34]. Previous work reported that AA carried by chitosan as an edible coating layer significantly reduced firmness during refrigerated storage of plums fruits [32]. It was suggested that coating and AA application could reduce the main cell wall-degrading enzyme (such as polygalacturonase and pectin methylesterase) activity.

Enzymatic browning is one of the most prominent industrial problems in fresh-cut products due to the oxidation of phenolic compounds [3]. Our results indicated that browning increased by increasing the storage period (Figure 2B). Blanching showed the lowest browning index for overall storage periods compared to other treatments and control. A significantly lower browning index for the MSE + AA treatment was found during the whole storage period compared to the control. Although ascorbic acid as an anti-browning agent in fresh-cut vegetables has been reported previously [3], the effect is primarily due to the role of ascorbic acid as an oxygen scavenger that prevents oxidation by polyphenol oxidase [35].

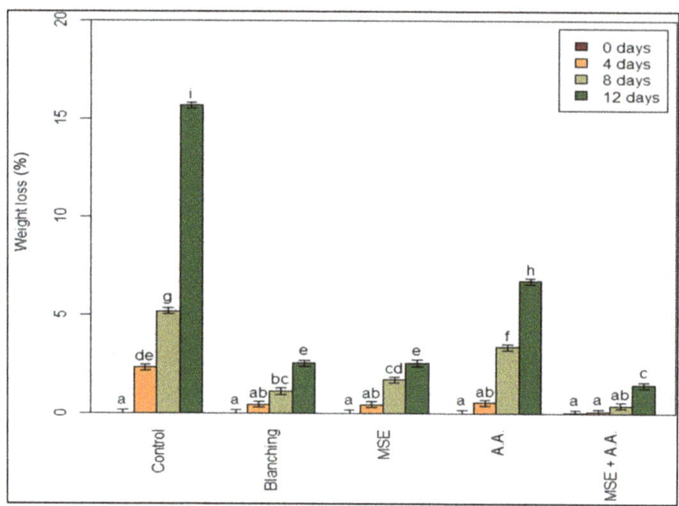

Figure 1. Effect of ascorbic acid (AA), mucilage seed extract (MSE), MSE + A.A, and blanching on weight loss (%) of fresh-cut potato strips. (Means with the same CLD letter are not statistically significantly different. Letters come from holm-corrected multiple comparisons across all times and treatments).

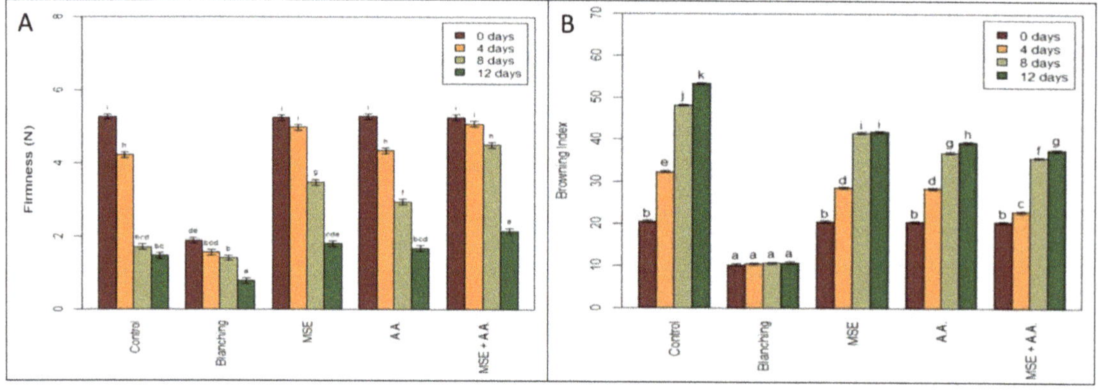

Figure 2. Effect of ascorbic acid (AA), mucilage seed extract (MSE), MSE + A.A, and blanching on firmness (N) (**A**) and browning index (**B**) of fresh-cut potato strips. (Means with the same CLD letter are not statistically significantly different. Letters come from a holm-corrected multiple comparisons, across all times and treatments).

3.2. *Total Phenolic Compound and Antioxidant Activity of Seedcake and MSE*

We determined the TPC in the dried garden cress seeds and in the seed mucilage after soaking in water to define the effect of the mucilage extraction method on the TPC content and to be sure of the quality of the garden cress that we used in this study. The aqueous extract contained 0.044 g GAE L^{-1}, and seeds contained 0.441 g GAE kg^{-1} dry weight of TPC. Several studies have previously determined the TPC of *Lepidium sativum* seeds. Zia-Ul-Haq et al. [36] reported that the aqueous extracts of *Lepidium sativum* seeds contain 0.0120 g GAE kg^{-1}. In addition, Chatoui et al. [17] reported that the TPC of *Lepidium sativum* seeds' aqueous extract contains 0.62 g GAE kg^{-1} of seed extract.

Meanwhile, the obtained results showed a lower content of TPC in *L. sativum* seed extract. The TPC content varies depending on plant variety, agronomic practices, seed collection stage, and climatic and area geological condition of where seeds are harvested [37]. Rafińska et al. [38] indicated that extracting TPC of dried *L. sativum* seeds with water was an effective method due to the lowest level of interfering substances with high molecular masses. In addition, high hydrophilicity characterizes the phenolic compounds present in the seeds, possibly due to several hydroxyl groups.

The mucilage seed extract of *L. sativum* showed good antioxidant activity determined with DPPH, which recorded 90%. Chatoui et al. [17] noticed that the increase in TPC increases the antioxidant activity of the extract.

3.3. Changes in Total Phenolic Content of Fresh-Cut Potato Stripes

For each treatment, the total phenolic content was determined in fresh-cut potato strips (Figure 3). The initial level of the phenolic compound ranged from 1827 to 3330 mg 100 g^{-1}; there was no difference between all treatments, except for the blanched sample, which recorded the lowest initial levels of TPC (0.1827 mg 100 g^{-1} of potato strips). The phenolic compounds' susceptibility to leaching from the plant tissue and degradation of heat-sensitive phenolic compounds can result in losses [39]. TPC contents of potato strips were significantly affected by the MSE treatment (Figure 3). MSE coating resulted in a 58% and 20% higher TPC content than the control sample (MSE + A.A.: 3291 mg 100 g^{-1}, MSE: 2499 mg 100 g^{-1} fresh weight of potato strips). The most significant effect was observed when MSE was supplemented with A.A., where the final retention of the TPC after 12 days of storage was 2315 mg 100 g^{-1} fresh weight of potato strips. These results are following El-Mogy et al. [3], who indicated that edible coating with *Cordia* gum (edible polysaccharide coating with a high content of phenolic compounds), when supplemented with AA, had a significant positive effect on TPC of artichoke bottoms during cold storage at 4 °C for 9 days. However, the TPC decreased during the storage period in fresh-cut potato strips, possibly due to oxidative processes [40]. The higher TPC of coated samples suggests lower phenol oxidation than the control treatment, which can be due to edible coating preventing contact between food and oxygen, as well as the degradation of phenolic compounds [41].

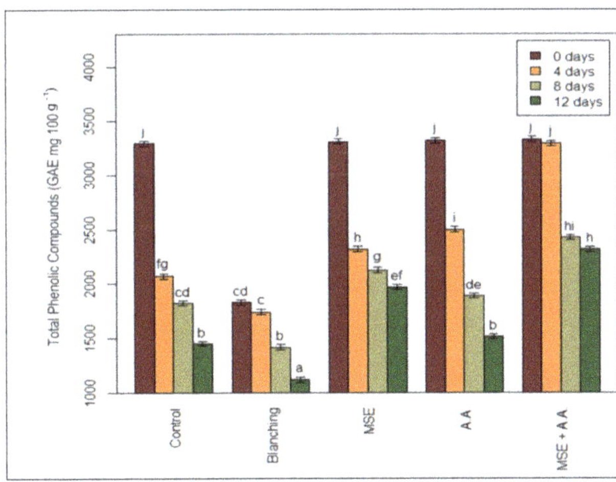

Figure 3. Effect of ascorbic acid (AA), mucilage seed extract (MSE), MSE + A.A, and blanching on total phenolic compounds (GAE mg 100 g^{-1}) of fresh-cut potato strips. (Means with the same CLD letter are not statistically significantly different. Letters come from holm-corrected multiple comparisons, across all times and treatments).

3.4. Oil Uptake (OU)%

Several studies have investigated the use of coating and blanching as a pre-treatment of frying for reducing the oil uptake [42,43]. All coating treatments (MSE and AA) significantly reduced OU% of fried potato strips compared to the uncoated sample (control, blanching) (Figure 4). As expected, the effect of coating showed that minimum oil content was related to MSE and MSE + AA (4.12, 4.08 g oil 100 g^{-1} potato), respectively.

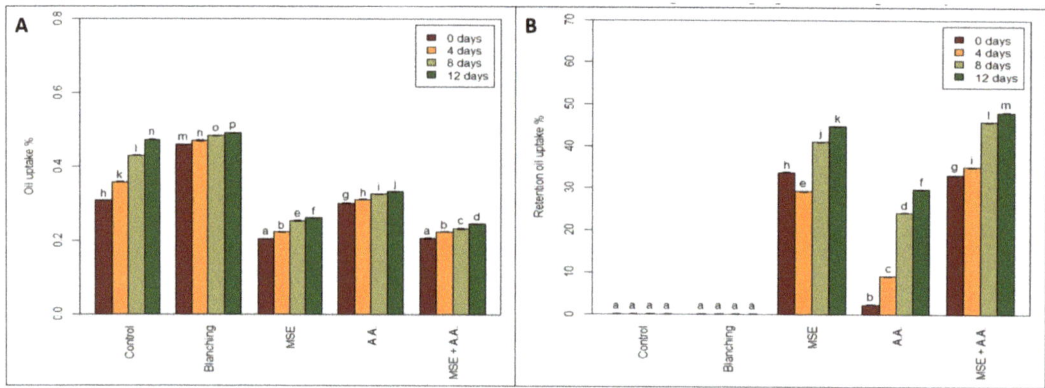

Figure 4. Effect of ascorbic acid (AA), mucilage seed extract (MSE), MSE + A.A, and blanching on oil uptake (**A**) and retention oil uptake (%) (**B**) of fresh-cut potato strips. (Means with the same CLD letter are not statistically significantly different. Letters come from holm-corrected multiple comparisons across all times and treatments.

In addition, the highest fat content was related to the blanched sample and control sample (non-coated) (9.19 and 6.16 g oil 100 g^{-1} potato, respectively). Previous studies have reported similar effects of blanching on oil uptakes [44,45]. Pedreschi and Moyano [45] noticed that blanching for high temperatures and short times (e.g., 97 °C, 2 min) before frying potato strips resulted in higher oil uptake than in control potato strips. This is because blanching involves the combined application of heat and water, which gelatinizes the starch on the surface of potato strips. This increase in oil content is undesirable for the acceptance of the product by the consumer.

The most important characteristics of edible coatings are oil barrier properties and flexibility because the volume of the food sample frequently changes during frying, and coating integrity may be compromised [42]. Consequently, hydrocolloidal coatings are used in various food applications, including adhesion, film shaping, thermal gelling, and non-charging properties. The form of the film properties of these hydrocolloids has stopped oil from being absorbed and has helped preserve the natural moisture content of the food. This may be the explanation for the deep frying of fried products using these hydrocolloids [46].

Therefore, the most effective coating treatments to reduce the percentage of oil uptake of fried potatoes were MSE + AA and MSE (33.08, 33.08%) at zero time. Moreover, there was no significant difference between MSE + AA and MSE at zero time or during the storage period of potato samples on oil content. The obtained results agreed with Garcia et al. [42], who found that the oil content of fried potatoes coated with 1% methylcellulose (MC) and 0.5% sorbitol was reduced in the range of 35–40%. This is related to the effective barrier properties of coating, which reduce oil uptake of fried potatoes. In addition, Mallikarjunan et al. [47] reported a protective layer formed on the surface of the fried products coated with cellulose derivatives during the initial stages of frying due to thermally induced gelation above 60°C. This layer could inhibit the transfer of fat and moisture between the sample and the frying oil.

To explain oil absorption, two main mechanisms are proposed as follows: condensation and capillary mechanisms; in both, inside the product, the oil penetrates through the pores. Therefore, the reduction in pore size and quantity can be due to the coating barrier limiting the oil uptake after frying. Thermo-gelling such as carboxymethyl cellulose (CMC), xanthan gum, and guar gum could lead to a more robust coating with low capillary pressures [13]. In addition, it is known that hydrocolloid treatment may alter the water-holding capacity and, consequently, prevent moisture replacement by oil [44].

3.5. Microbiological Analysis

Figure 5 showed that the microbial load of the fresh-cut potato strips was affected by the cold storage duration at 5 °C for 12 days. All treatments were free from *E. coli* until the end of storage and had the same total microbial count, which ranged from 2.01 to 2.06 log CFU g^{-1} at zero time, except for the control, which had the highest count 3.02 log CFU g^{-1}. At the end of storage, the lowest microbial count (2.91, 2.92 log CFU g^{-1}) was found in both the MSE + AA and blanched samples, whereas the highest count (5.40 log CFU g^{-1}) was observed in the control after 12 days of storage. In addition, at zero-time, the mold and yeast counts contained in all potato samples were below the detection limit (15 CFU g^{-1}), except for the control and AA samples (2.81 and 1.69 log CFU g^{-1}, respectively). The superior samples were MSE and MSE + AA, which remained free from mold and yeast growth after up to 8 d of cold storage.

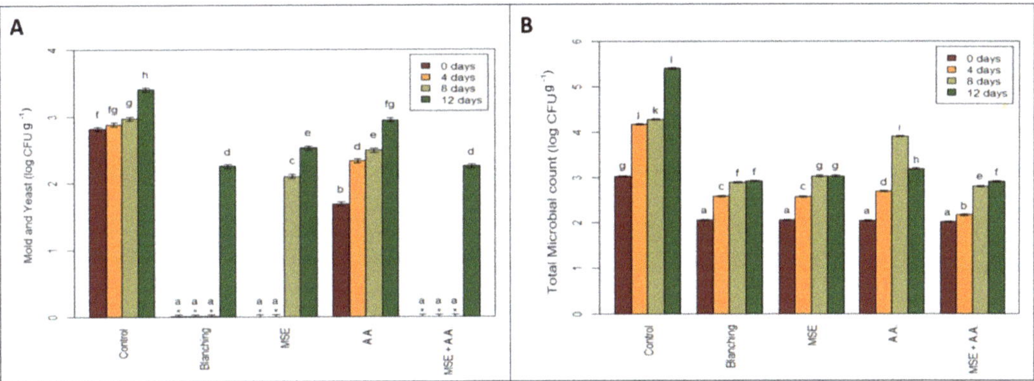

Figure 5. Effect of ascorbic acid (AA), mucilage seed extract (MSE), MSE + A.A, and blanching on mold and yeast (log CFU g^{-1}) (**A**) and total microbial count (log CFU g^{-1}) (**B**) of fresh-cut potato strips. (Means with the same CLD letter are not statistically significantly different. Letters come from holm-corrected multiple comparisons across all times and treatments.).

There was no significant difference between all mold and yeast count treatments at the end of storage except for the control sample. On day 12, the control presented the highest count (3.40 log CFU g^{-1}) while all other samples presented a lower count ranging from 2.25 to 2.95 log CFU g^{-1}.

It is understood that microbiological safety is the most crucial factor in storing fresh-cut fruit and vegetables. Therefore, senescence and maturity reduce firmness, making the product more vulnerable to microbial attack [48]. In fresh-cut vegetables, both fungi and bacteria are significant causes of spoilage. The high-water activity of most vegetables and close neutral pH renders them adequate hosts for all kinds of microorganisms. However, the faster bacterial growth rate typically helps them to compete with the fungi more efficiently [49,50]. Cacace et al. [51] observed a more minor increase in the microbial count of fresh-cut potato treated with different chemicals such as erythorbic acid (5%) and citric acid (1%) with storage at 5 °C and reported a decrease in microbial growth with an increase in acidity. Dipping treatment using organic acids like ascorbic acid can also possess bactericidal properties [52]. The antimicrobial activity of organic acids is due to a decrease in environmental pH, disturbance of membrane transport and permeability, accumulation of anions, or a decrease in internal cellular pH due to the dissociation of acid from hydrogen ions.

Similarly, Licciardello et al. [53] showed that locust bean gum-based edible coating effectively reduced microbial growth. In addition, Barzegar et al. [54] noticed that *Lepidium sativum* seed mucilage-based edible coating containing 1.5% *Heracleum lasiopetalum* essential oil resulted in a significant increase in microbiological stability of the beef samples stored at 4 °C for 9 days, compared to the control. Such edible coating could prevent contact between food and oxygen. Therefore, incorporating mucilage of seeds such as *Lepidium sativum* can improve the barrier properties of the edible films against the transport of moisture and oxygen. Moreover, they can extend the shelf life of final products with their content of bioactive components (phenolic compounds), which have an antimicrobial effect [55]. However, the total microbial count and mold and yeast during cold storage is still well below the critical limits [56]. They mention that the critical limits for total microbial count and mold and yeast of vegetable were 108 CFU g^{-1} and 105 CFU g^{-1}, respectively.

3.6. Sensory Evaluation

The most essential factors for consumer demand are sensory quality and nutrient contents of processed food products [57]. The ANOVA results showed no difference in the sensory appraisal of color, taste, odor, texture, and overall appearance of the samples among the different treatments after frying. However, before frying, the treatments showed an effect on the color and texture of the fresh potato strips, where the color of the control samples had the least rating, whereas all the treatments had no difference. For texture, blanched (before frying) samples had the least acceptability, and all other treatments, including the control, were the same. The color differences were related to the enzymatic and nonenzymatic browning (oxidation of phenolic compounds). It was expected that MSE might alter the taste of fried potato strips due to its pungent flavor [17]. However, the taste scores were highest for MSE samples. In the case of the blanched sample, a low score of taste and texture (of fresh and fried) regarding its high content of oil, as mentioned before, and heat treatment, which reduces firmness of the tissue. Reis et al. [24] reported that hydrocolloid coatings (CMC and xanthan gum) were used to improve the sensory attributes such as taste, texture, color, and appearance of French fries.

A non-metric multidimensional (NMDS) scaling for sensory parameters using mono-MDS is presented in Figure 6. Additionally, circular points are replicates, and solid squares are centroids for the data ellipses by treatment. In comparison, Figure 6B is an arrow plot, which shows that MSE and MSE + AA centroids lie at a higher level in the multidimensional spaces of the treatments, which means that the samples with MSE coatings showed higher rating results in the sensory evaluation. In Figure 6B, the weighing of each sensory attributes was calculated by the ordination techniques. The weightings for that attribute are further apart and the arrow from the center points higher, which shows that the total appearance before frying, color after frying, and odor after frying had the highest impact on the overall sensory evaluation results.

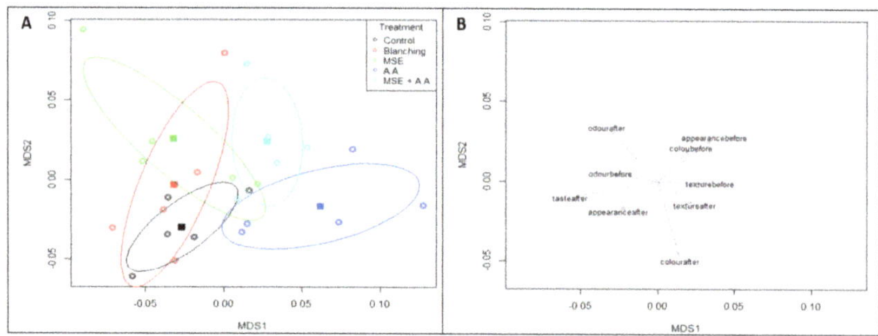

Figure 6. Non-metric multidimensional (NMDS) scaling for sensory parameters using mono-MDS. Circular points are replicates, solid square points are centroids for the data ellipses (by treatment) (**A**). Arrow plots showing weightings for sensory scores (**B**).

4. Conclusions

The novel method of utilizing garden cress seed extract into an edible coating to incorporate ascorbic acid help reduced the weight loss, browning index and preserve the firmness and total phenolic compound during storage of fresh-cut potatoes. Moreover, the coating helped reduce the oil uptake of the potato strips after frying. The sensory evaluation results showed that the sensory perception of the fried potato fries was enhanced using these edible coatings. Further research in commercial production and application of the garden cress seed extract is required for industrial application and testing its applicability to other fresh fruits and vegetable crops.

Author Contributions: Conceptualization, M.R.A. and M.M.E.-M.; methodology, A.A.E.-Y. and M.M.E.-M.; software, A.P., G.N. and T.W.; validation, G.N., H.G.A.E.-G. and N.E.N.; formal analysis, A.P., T.W. and M.F.M.I.; investigation, M.R.A. and M.M.E.-M.; resources, N.E.N., A.A.E.-Y. and M.F.M.I.; data curation, H.G.A.E.-G.; writing—original draft preparation, M.R.A. and M.M.E.-M.; writing—review and editing, M.R.A., A.P., G.N. and T.W.; visualization, A.P. and H.G.A.E.-G.; supervision, M.M.E.-M.; project administration, M.R.A.; funding acquisition, M.R.A., N.E.N. and M.F.M.I. All authors have read and agreed to the published version of the manuscript.

Funding: This research was funded by Cairo University, Faculty of Agriculture, Giza, Egypt.

Institutional Review Board Statement: Not applicable.

Informed Consent Statement: Not applicable.

Data Availability Statement: Not applicable.

Acknowledgments: The authors would like to thank Stephen Young, a statistician at NRI, for his help in the data analysis.

Conflicts of Interest: The authors confirm that there is no conflict of interest.

References

1. Abuarab, M.E.; El-Mogy, M.M.; Hassan, A.M.; Abdeldaym, E.A.; Abdelkader, N.H.; El-Sawy, M. The effects of root aeration and different soil conditioners on the nutritional values, yield, and water productivity of potato in clay loam soil. *Agronomy* **2019**, *9*, 418. [CrossRef]
2. Tian, J.; Chen, J.; Ye, X.; Chen, S. Health benefits of the potato affected by domestic cooking: A review. *Food Chem.* **2016**, *202*, 165–175. [CrossRef]
3. El-Mogy, M.M.; Parmar, A.; Ali, M.R.; Abdel-Aziz, M.E.; Abdeldaym, E.A. Improving postharvest storage of fresh artichoke bottoms by an edible coating of *Cordia myxa* gum. *Postharvest Biol. Technol.* **2020**, *163*, 111143. [CrossRef]
4. Awad, A.H.R.; Parmar, A.; Ali, M.R.; El-Mogy, M.M.; Abdelgawad, K.F. Extending the shelf-life of fresh-cut green bean pods by ethanol, ascorbic acid, and essential oils. *Foods* **2021**, *10*, 1103. [CrossRef]
5. Liu, X.; Lu, Y.; Yang, Q.; Yang, H.; Li, Y.; Zhou, B.; Li, T.; Gao, Y.; Qiao, L. Cod peptides inhibit browning in fresh-cut potato slices: A potential anti-browning agent of random peptides for regulating food properties. *Postharvest Biol. Technol.* **2018**, *146*, 36–42. [CrossRef]
6. Hou, Z.; Feng, Y.; Wei, S.; Wang, Q. Effects of curing treatment on the browning of fresh-cut potatoes. *Am. J. Potato Res.* **2014**, *91*, 655–662. [CrossRef]
7. Tsouvaltzis, P.; Deltsidis., A.; Brecht, J.K. Hot water treatment and pre-processing storage reduce browning development in fresh-cut potato slices. *Hort. Sci.* **2011**, *46*, 1282–1286. [CrossRef]
8. Yeoh, W.K.; Ali, A. Ultrasound treatment on phenolic metabolism and antioxidant capacity of fresh-cut pineapple during cold storage. *Food Chem.* **2017**, *216*, 247–253. [CrossRef] [PubMed]
9. Gao, H.; Zeng, Q.; Ren, Z.; Li, P.; Xu, X. Effect of exogenous γ-aminobutyric acid treatment on the enzymatic browning of fresh-cut potato during storage. *J. Food Sci. Technol.* **2018**, *55*, 5035–5044. [CrossRef] [PubMed]
10. Ru, X.; Tao, N.; Feng, Y.; Li, Q.; Wang, Q. A novel anti-browning agent 3-mercapto-2-butanol for inhibition of fresh-cut potato browning. *Postharvest Biol. Technol.* **2020**, *170*, 111324. [CrossRef]
11. Yoruk, R.; Marshall, M.R. A Survey on the Potential Mode of Inhibition for Oxalic Acid on Polyphenol Oxidase. *J. Food Sci.* **2003**, *68*, 2479–2485. [CrossRef]
12. Landi, M.; Degl'Innocenti, E.; Guglielminetti, L.; Guidi, L. Role of ascorbic acid in the inhibition of polyphenol oxidase and the prevention of browning in different browning-sensitive *Lactuca sativa* var. capitata (L.) and *Eruca sativa* (Mill.) stored as fresh-cut produce. *J. Sci. Food Agric.* **2013**, *93*, 1814–1819. [CrossRef]
13. Garmakhany, A.D.; Mirzaei, H.O.; Nejad, M.K.; Maghsudlo, Y. Study of oil uptake and some quality attributes of potato chips affected by hydrocolloids. *Eur. J. Lipid Sci. Technol.* **2008**, *110*, 1045–1049. [CrossRef]

14. Ali, M.R.; Mohamed, R.M., Abedelmaksoud, T.G. Functional strawberry and red beetroot jelly candies rich in fibers and phenolic compounds. *Food Syst.* **2021**, *4*, 12–18. [CrossRef]
15. Krokida, M.K.; Oreopoulou, V.; Maroulis, Z.B.; Marinos-Kouris, D. Effect of pre-drying on quality of french fries. *J. Food Eng.* **2021**, *49*, 347–354. [CrossRef]
16. Mohite, S.Y.; Gharal, D.B.; Ranveer, R.C.; Sahoo, A.K.; Ghosh, J.S. Development of health drink enriched with processed garden cress seeds. *Am. J. Food Technol.* **2012**, *7*, 571–576. [CrossRef]
17. Chatoui, K.; Harhar, H.; El Kamli, T.; Tabyaoui, M. Chemical composition and antioxidant capacity of *Lepidium sativum* seeds from four regions of Morocco. *Evid. Based Complement. Altern. Med.* **2020**, 7302727. [CrossRef] [PubMed]
18. Baregma, C.; Goyal, A. Phytoconstituents activity and medicinal use of *Lepidium Sativum* Linn.: A review. *Asian J. Pharm. Clin. Res.* **2019**, *12*, 45–50. [CrossRef]
19. Jouk, M.; Khazaeib, N.; Ghasemlouc, M.; HadiNezhad, M. Effect of glycerol concentration on edible film production from cress seed carbohydrate gum. *Carbohydr. Polym.* **2013**, *96*, 39–46. [CrossRef]
20. Salehi, F. Characterization of New Biodegradable Edible Films and Coatings Based on Seeds Gum: A Review. *J. Packag. Technol. Res.* **2019**, *3*, 193–201. [CrossRef]
21. Karamkhani, M.; Anvar, S.A.A.; Ataee, M. The use of active edible coatings made from a combination of *Lepidium sativum* gum and carvacrol to increase shelf life of farmed shrimp kept under refrigerator condition. *Iran. J. Aquat. Anim. Health* **2018**, *4*, 55–72. [CrossRef]
22. Rizzo, V.; Muratore, G. The application of essential oils in edible coating: Case of study on two fresh cut products. *Int. J. Clin. Nutr. Diet* **2020**, *6*, 149. [CrossRef]
23. Karazhiyan, H.; Razavi, S.M.A.; Phillips, G.O. Extraction optimization of a hydrocolloid extract from cress seed (*Lepidium sativum*) using response surface methodology. *Food Hydrocoll.* **2011**, *25*, 915–920. [CrossRef]
24. Ali, M.R.; El Said, R.M. Assessment of the potential of Arabic gum as an antimicrobial and antioxidant agent in developing vegan "egg-free" mayonnaise. *J. Food Saf.* **2020**, *40*, e12771. [CrossRef]
25. Reis, F.R.; Masson, M.L.; Waszcznskyjet, N. Influence of a blanching pre-treatment on color, oil uptake and water activity of potato activity of potato sticks, and its optimization. *J. Food Process Eng.* **2008**, *31*, 833–852. [CrossRef]
26. El-Mogy, M.M.; Ali, M.R.; Darwish, O.S.; Rogers, H.J. Impact of salicylic acid, abscisic acid, and methyl jasmonate on postharvest quality and bioactive compounds of cultivated strawberry fruit. *J. Berry Res.* **2019**, *9*, 333–348. [CrossRef]
27. El-Mogy, M.M.; Garchery, C.; Stevens, R. Irrigation with salt water affects growth, yield, fruit quality, storability and marker-gene expression in cherry tomato. *Acta Agric. Scand. Sect. B Soil Plant Sci.* **2018**, *68*, 727–737. [CrossRef]
28. Bligh, E.G.; Dyer, W.J. A rapid method of total lipid extraction and purification. *Can. J. Biochem. Physiol.* **1959**, *37*, 911–917. [CrossRef]
29. Shehata, S.A.; Abdeldaym, E.A.; Ali, M.R.; Mohamed, R.M.; Bob, R.I.; AbdelGawad, K.F. Effect of Some Citrus Essential Oils on Post-Harvest Shelf Life and Physicochemical Quality of Strawberries during Cold Storage. *Agronomy* **2020**, *10*, 1466. [CrossRef]
30. ISO 16654. Horizontal method for the detection of Escherichia coli 0157 2001. Available online: https://www.iso.org/standard/29821.html (accessed on 30 June 2021).
31. Ngobese, N.Z.; Workneh, T.S.; Siwela, M. Effect of low-temperature long-time and high-temperature short-time blanching and frying treatments on the French fry quality of six Irish potato cultivars. *J. Food Sci. Technol.* **2017**, *54*, 507–517. [CrossRef] [PubMed]
32. Liu, K.; Yuan, C.; Chen, Y.; Li, H.; Liu, J. Combined effects of ascorbic acid and chitosan on the quality maintenance and shelf life of plums. *Sci. Hortic.* **2014**, *176*, 45–53. [CrossRef]
33. Noctor, G.; Foyer, C.H. Ascorbate and glutathione: Keeping active oxygen under control. *Annu. Rev. Plant Physiol. Plant Mol. Biol.* **1998**, *49*, 249–279. [CrossRef]
34. El-Mogy, M.M.; Ludlow, R.A.; Roberts, C.; Müller, C.T.; Rogers, H.J. Postharvest exogenous melatonin treatment of strawberry reduces postharvest spoilage but affects components of the aroma profile. *J. Berry Res.* **2019**, *9*, 297–307. [CrossRef]
35. Rico, D.; Martín-Diana, A.B.; Barat, J.M.; Barry-Ryan, C. Extending and measuring the quality of fresh-cut fruit and vegetables: A review. *Trends Food Sci. Technol.* **2007**, *18*, 373–386. [CrossRef]
36. Zia-Ul-Haq, M.; Ahmad, S.; Calani., L.; Mazzeo, T.; Del Rio, D.; Pellegrini, N.; De Feo, V. Compositional study and antioxidant potential of *Ipomoea hederacea* Jacq. and *Lepidium sativum* L. seeds. *Molecules* **2012**, *17*, 10306–10321. [CrossRef] [PubMed]
37. Zia-Ul-Haq, M.; Shahid, S.A.; Ahmad, S.; Qayum, M.; Khan, I. Antioxidant potential of various parts of *Ferula assafoetida* L. *J. Med. Plant Res.* **2012**, *6*, 3254–3258. [CrossRef]
38. Rafińska, K.; Pomastowski, P.; Rudnicka, J.; Krakowska, A.; Maruśka, A.; Narkute, M. and Buszewski, B. Effect of solvent and extraction technique on composition and biological activity of *Lepidium sativum* extracts. *Food Chem.* **2019**, *289*, 16–25. [CrossRef]
39. Chellaram, C.; Parthasarathy, V.; Praveen, M.M.; John, A.A.; Anand, T.P.; Priya, G.; Kesavan, D. Analysis of phenolic content and antioxidant capacity of potato, *Solanum Tuberosum* L from Tamilnadu Region, India. *APCBEE Procedia* **2014**, *8*, 105–108. [CrossRef]
40. Coelho, D.G.; Andrade, M.T.D.; Mélo Neto, D.F.D.; Ferreira-silva, S.L.; Simões, A.N.D. Application of antioxidants and edible starch coating to reduce browning of minimally-processed cassava. *Rev. Caatinga* **2017**, *30*, 503–512. [CrossRef]
41. Tomás-Barberan, F.A.; Espín, J.C. Phenolic compounds and related enzymes as determinants of quality in fruits and vegetables. *J. Sci. Food Agric.* **2001**, *81*, 853–876. [CrossRef]
42. García, M.A.; Ferrero, C.; Campana, A.; Bértola, N.; Martino, M.; Zaritzky, N. Methylcellulose Coatings Applied to Reduce Oil Uptake in Fried Products. *Food Sci. Technol. Int.* **2004**, *10*, 339–346. [CrossRef]

43. Moyano, P.C.; Pedreschi, F. Kinetics of oil uptake during frying of potato slices: Effect of pre-treatments. *LWT Food Sci. Technol.* **2006**, *39*, 285–291. [CrossRef]
44. Rimac-Brnčić, S.; Lelas, V.; Rade, D.; Šimundić, B. Decreasing of oil absorption in potato strips during deep fat frying. *J. Food Eng.* **2004**, *64*, 237–241. [CrossRef]
45. Pedreschi, F.; Moyano, P. Effect of pre-drying on texture and oil uptake of potato chips. *LWT Food Sci. Technol.* **2005**, *38*, 599–604. [CrossRef]
46. Sakhale, B.K.; Badgujar, J.B.; Pawar, V.D.; Sananse, S.L. Effect of hydrocolloids incorporation in casing of samosa on reduction of oil uptake. *J. Food Sci. Technol.* **2011**, *48*, 69–772. [CrossRef] [PubMed]
47. Mallikarjunan, P.; Chinnan, M.S.; Balasubramaniam, V.M.; Phillips, R.D. Edible Coatings for Deep-fat Frying of Starchy Products. *LWT Food Sci. Technol.* **1997**, *30*, 709–714. [CrossRef]
48. Silveira, A.C.; Oyarzún, D.; Sepúlveda, A.; Escalona, V. Effect of genotype, raw-material storage time and cut type on native potato suitability for fresh-cut elaboration. *Postharvest Biol. Technol.* **2017**, *128*, 1–10. [CrossRef]
49. Brackett, R.E. Microbiological Spoilage and Pathogens in Minimally Processed Refrigerated Fruits and Vegetables. In *Minimally Processed Refrigerated Fruits & Vegetables*; Wiley, R.C., Ed.; Springer: Boston, MA, USA, 1994; pp. 269–312.
50. Alam, M.; Ahlström, C.; Burleigh, S.; Olsson, C.; Ahrné, S.; El-Mogy, M.M.; Molin, G.; Jensén, P.; Hultberg, M.; Alsanius, B.W. Prevalence of *Escherichia coli* O157:H7 on spinach and rocket as affected by inoculum and time to harvest. *Sci. Hortic.* **2014**, *165*, 235–241. [CrossRef]
51. Cacace, J.E.; Delaquis, P.J.; Mazza, G. Effect of chemical inhibitors and storage temperature on the quality of fresh-cut potatoes. *J. Food Qual.* **2002**, *25*, 181–195. [CrossRef]
52. Bico, S.L.S.; Raposo, M.F.J.; Morais, R.M.S.C.; Morais, A.M.M.B. Combined effects of chemical dip and/or carrageenan coating and/or controlled atmosphere on quality of fresh-cut banana. *Food Control* **2009**, *20*, 508–514. [CrossRef]
53. Licciardello, F.; Lombardo, S.; Rizzo, V.; Pitino, I.; Pandino, G.; Strano, M.G.; Muratore, G.; Restuccia, C.; Mauromicale, G. Integrated agronomical and technological approach for the quality maintenance of ready-to-fry potato sticks during refrigerated storage. *Postharvest Biol. Technol.* **2018**, *136*, 23–30. [CrossRef]
54. Barzegar, H.A.; Behbahani, B.; Mehrnia, M.A. Quality retention and shelf life extension of fresh beef using Lepidium sativum seed mucilage-based edible coating containing *Heracleum lasiopetalum* essential oil: An experimental and modeling study. *Food Sci. Biotechnol.* **2020**, *29*, 717–728. [CrossRef] [PubMed]
55. Beikzadeh, S.; Khezerlou, A.; Jafari, S.M.; Pilevar, Z.; Mortazavian, A.M. Seed mucilages as the functional ingredients for biodegradable films and edible coatings in the food industry. *Adv. Colloid Interface Sci.* **2020**, *280*, 102164. [CrossRef]
56. Jacxsens, L.; Devlieghere, F.; Debevere, J. Predictive modelling for packaging design: Equilibrium modified atmosphere packages of fresh-cut vegetables subjected to a simulated distribution chain. *Int. J. Food Microbiol.* **2002**, *73*, 331–341. [CrossRef]
57. Łaska-Zieja, B.; Marcinkowski, D.; Golimowski, W.; Niedbała, G.; Wojciechowska, E. Low-Cost Investment with High Quality Performance. Bleaching Earths for Phosphorus Reduction in the Low-Temperature Bleaching Process of Rapeseed Oil. *Foods* **2020**, *9*, 603. [CrossRef] [PubMed]

Article

Effects of Light Exposure, Bottle Colour and Storage Temperature on the Quality of *Malvasia delle Lipari* Sweet Wine

Elena Arena [1], Valeria Rizzo [1,*], Fabio Licciardello [2], Biagio Fallico [1] and Giuseppe Muratore [1]

[1] Department of Agriculture, Food and Environment (Di3A), University of Catania, Via Santa Sofia, 100, 95123 Catania, Italy; elena.arena@unict.it (E.A.); biagio.fallico@unict.it (B.F.); g.muratore@unict.it (G.M.)

[2] Department of Life Sciences, University of Modena and Reggio Emilia, Via Amendola 2, 42122 Reggio Emilia, Italy; fabio.licciardello@unimore.it

* Correspondence: vrizzo@unict.it

Abstract: The influence of light exposure, bottle color and storage temperature on the quality parameters of *Malvasia delle Lipari* (MdL) sweet wine were investigated. Wine samples bottled in clear-colored (colorless, green and amber) glass were stored under different artificial lighting conditions, in order to simulate the retail environment (one cool-white, fluorescent lamp) and to perform an accelerated test (four and six cool-white, fluorescent lamps). The storage temperature was kept constant (25 °C) for the first 90 days of the experiment and then samples were monitored for up to 180 days at higher temperatures (30, 35 and 40 °C). The principal enological parameters, total phenols, color, 5-hydroxymethylfurfural (HMF) and 2-furaldehyde (2F) contents were studied. The shelf-life test pointed out minimum variations of the basic chemical parameters, while the quality attributes most affected by lighting were color, together with HMF and 2F levels which, hence, can be considered as indicators of the severity of storage conditions.

Keywords: Malvasia; sweet wine; shelf-life; accelerated shelf-life test; 5-hydroxymethylfurfural; 2-furaldehyde

Citation: Arena, E.; Rizzo, V.; Licciardello, F.; Fallico, B.; Muratore, G. Effects of Light Exposure, Bottle Colour and Storage Temperature on the Quality of *Malvasia delle Lipari* Sweet Wine. *Foods* **2021**, *10*, 1881. https://doi.org/10.3390/foods10081881

Academic Editor: Maurizio Ciani

Received: 20 June 2021
Accepted: 12 August 2021
Published: 14 August 2021

1. Introduction

The primary objective of packaging is to protect and retain, as much as possible, the quality of foods and beverages. The classical packaging material for wine is glass, appreciated primarily for its oxygen barrier [1], clarity and inertness: this feature, with respect to the migration of low molecular weight compounds from the package to the product and/or flavor scalping by the packaging material is of the utmost importance [2,3]. Wine has traditionally been stored in glass bottles of different colors and shapes, whose selection is often driven from market forces in the attempt to make the wine readily identifiable and more attractive to the consumer.

Wine shelf-life is defined as the time that it remains stable, in terms of its chemistry, microbiology and biochemistry as well as its sensory properties [4], however chemical changes may occur during storage, improving some quality parameters. Color is among the main sensory attributes determining consumers' preference and it is considered a major feature for the assessment of red wine quality [5]. The color of white wines tends to brown after bottling [6], due to the effect of oxidation of the phenolic compound and to enzymatic reactions [7,8]. Thus, during storage and ageing, the chemical composition of wine is subjected to continuous changes, which may be desired or not, depending on the type of wine. Phenolics and volatile compounds as well as color and nutrients could be influenced by the lighting and temperature of the retail shops, and the optimization of packaging variables could contribute to wine quality preservation, from production to consumption.

Many authors have studied the effects of time, temperature and storage conditions on the phenolics composition and color of red wines [9–11]. A significant decrease of phenolics content and a change in the color of the white wine from pale yellow to yellow brown during storage, was reported. Time of storage seems to influence the color parameters and the total phenols content more with respect to temperature and light exposure [12]. Other authors [13] did not find color change, and total phenols decreased up to nine months of storage of Hellenic varietal white wine.

The deteriorative effect of light depends on other factors, such as the duration of light exposure, light spectra and intensity and the composition of the food product. It is well known that the exposure of some foods to ultraviolet radiation and visible light accelerates oxidative deterioration. In alcoholic beverages, the aroma, flavor and color easily deteriorate for light-induced oxidative processes. The effect of light can be explained by both photolytic autoxidation, which corresponds to the production of radicals during exposure to UV light, and photosensitized oxidation, that occurs in the presence of photosensitizers and ultraviolet or visible light. The roles of light, temperature and, more frequently, also the color of the glass bottle in white wine, have been extensively analyzed in order to evaluate their impact on light-induced oxidative degradation, changes on pigment, phenolic compounds, and mainly on enological parameters [14–18]. Clark et al. [19] discussed the specific determination of the critical wavelengths for photoactivation, leading to the formation of glyoxylic acid, considering the capacity of two types of glass bottles (clear and dark green) to protect or limit the photodegradation of tartaric acid to glyoxylic acid. It is well known that darker colored bottles tend to give greater protection to wine from the influences of light exposure on the assumption that dark colors do not allow the transmission of UV radiation. More recently, environmental groups have proposed that the wine industry should move away from dark green to flint or amber glass bottles as a contribution towards reducing recycling costs and energy demands [17].

The *Malvasia delle Lipari DOC* wine (MdL) produced in the Aeolian Islands (Italy) as a liqueur wine, with minimum developed alcoholic strength of 20% v/v, is one of the most ancient and aromatic wines of Sicily. The art. 6 of the Regulation D. P. R. 20/09/1973 "Assignment of the denomination of controlled origin of "Malvasia delle Lipari" wine" [20] describes the main characteristics of the MdL wine when it is released for consumption. The attention is also focused on the color, which must be golden yellow or amber yellow, underlining the importance of such a factor. Considering how the color could change during the storage, we focused on different colored glass bottles and different storage conditions during the shelf-life of the product.

According to production techniques, the grapes (Malvasia delle Lipari and Corinto Nero 95% and 5% respectively, as defined in D.P.R. 20/9/1973 [20]) are gathered when they are fully ripe and then sun-dried for 10–15 days on large mats made of bamboo canes, to reduce moisture and increase the sugar level (up to 32%) to obtain a much more aromatic white wine [21].

In MdL, the high level of sugars and the low pH value suggest the possibility to develop sugar degradation products, such as 5-hydroxymethylfurfural (HMF) and 2-furaldehyde (2F), generally considered as indicators of heat processing and/or the prolonged storage of foods [21].

Both HMF and 2F are the most important intermediate product of the acid-catalyzed degradation of hexose and pentose, respectively, and HMF derived also from the decomposition of 3-deoxyosone during the early stage of Maillard reaction [22]. HMF is used as a quality parameter in several processed foods [23–29]. In fortified wine with different sweetness levels, the concentrations of HMF were strongly dependent on time and temperature used during winemaking, and strictly related to the sugar content of the wine [30]. Moreover, both HMF and 2F are involved in the aroma of sweet, fortified white wines aged in oxygen-free conditions [31].

In recent years, several papers have debated on the safety of these compounds. HMF has been shown to be converted, in vitro and in vivo, into 5-sulfoxymethylfurfural, which has been reported to be cytotoxic, mutagenic and carcinogenic [32–34], and can be a poison for the nervous system [35].

The aim of this work was to study the influence of different lighting conditions, bottle color and storage temperature on the quality parameters of MdL wine: moreover, for the first time, the kinetics parameters, *k* and Ea, were determined for HMF and 2F formation in sweet wine.

2. Materials and Methods

2.1. Characterization of Coloured Wine Bottles

Clear glass bottles (200 mL) in three different colors (colorless, green and amber) were characterized in terms of light transmission properties, measuring on five glass samples exposed to a cool-white, fluorescent lamp (Osram Lumilux® 36W/840, Munich, Germany) by a digital handle photo-radiometer (Delta OHM HD 2102.2, Delta OHM S.r.l., Caselle di Selvazzano (PD), Italy) equipped with probes for the measurement of illuminance (lux) (400–800 nm) and irradiance (W m^{-2}) in the UVA (315–400 nm) and UVB (280–315 nm) regions. The light shielding effect offered by the packaging materials, expressed as percentage of irradiance, was calculated as:

$$\text{Light shielding} = I_{sample}/I_{air} \times 100 \quad (1)$$

where I_{sample}: irradiance recorded in the presence of sample; I_{air}: irradiance recorded in the absence of sample [36].

2.2. Sampling

Malvasia delle Lipari DOC wine (MdL) produced in Sicily was kindly provided by a local manufacturer. The analytical parameters (pH, °Brix, total acidity, volatile acidity, alcoholic strength, total phenols, CIE Lab color parameters, HMF and 2F) were determined on the wine as received, before dispensing into the previously described clear-colored glass bottles (defined as Time "0").

Bottles were fulfilled (200 mL with 7 mL headspace), then stoppered with crown corks and stored horizontally up to six months: three months under different lighting conditions and three months under different temperature conditions. Overall, 144 bottles were considered for the study of shelf-life, according to the experimental plan reported in Table 1.

Table 1. Experimental design.

Storage Conditions		Bottle Color	Storage Time
CWF Lamp	**Temperature**		
n = 4	*n* = 4	*n* = 3	*n* = 6
0 (control)	25 °C (control)	Colorless	30
1	30	Green	60
4	35	Amber	90
6	40		120
			150
			180

During the first three months, the storage temperature was 25 °C. Samples were divided into three batches and exposed to different lighting conditions: (a) under constant illumination produced by one cool-white, fluorescent (CWF) lamp (Osram Lumilux® 36W/840, Munich, Germany), as to simulate the retailers conditions; (b) under constant illumination produced by four CWF lamps and (c) under constant illumination produced by six CWF lamps, in order to perform an accelerated test, by exposing the sweet wines to extreme conditions. Lamps were placed 30 cm above samples. Table 2 reports the illuminance (lux) and UVA and UVB irradiance (W m^{-2}) values recorded for each selected lighting condition.

Table 2. Measurements of illuminance (lux) and irradiance in the UVA and UVB regions (W m^{-2}) recorded by digital handle photo-radiometer (values are mean of five measurements).

Storage Condition	Illuminance (lux)	UVA (W m^{-2})	UVB (W m^{-2})
1 CWF lamp	2671 ± 9	101.4^{-3} ± 0.1	10.71^{-3} ± 0.03
4 CWF lamps	9028 ± 16	284.5^{-3} ± 0.6	31.53^{-3} ± 0.06
6 CWF lamps	16127 ± 13	478.2^{-3} ± 0.8	53.91^{-3} ± 0.07

After, for the next three months, MdL samples were stored under constant illumination, one cool-white, fluorescent (CWF) lamp, at three different temperatures: 30, 35 and 40 °C. Data loggers (Smart Reader SR04, ACR Systems Inc., Vancouver, BC, Canada) were used to monitor the storage temperatures throughout the trial.

MdL samples were periodically withdrawn at 30, 60, 90, 120, 150 and 180 days from being bottled and analyzed.

2.3. *Analytical Parameters*

Total acidity and volatile acidity, °Brix, pH and alcohol content were determined according to the OIV official methods [37]. Chromatic characteristics (CIE Lab) were determined according to Method OIV-MA-AS2-11 [37] using a spectrophotometer (Cary 1E, Varian, Leinì (TO) Italy) with a 1 cm quartz cell. The CIE parameters (L^*, a^*, b^*, C, h) were determined by the "Color Calculations" spectrophotometer software (Cary WinUV 1.3, Varian, Leinì (TO) Italy).

All analyses were carried out in duplicate, using chemicals (Sigma-Aldrich, Milan, Italy) and solvents of analytical grade (J. T. Baker, Deventer, The Netherlands) and HPLC grade (Merck, Milan, Italy).

2.4. *HMF and 2-Furaldehyde*

Aliquots of MdL sample were opportunely diluted with water (JT. Baker, Deventer, Holland), filtered through a 0.45-µm filter (Albet) and injected into an HPLC system (Shimadzu Class VP LC-10ADvp) equipped with a DAD (Shimadzu SPD-M10Avp, Kyoto, Japan). The column was a Gemini NX C18 (150 mm × 4.6 mm, 5 µm) (Phenomenex, Torrance, CA, USA), fitted with a guard cartridge packed with the same stationary phase. The HPLC conditions were the following: isocratic mobile phase, 90% water (Riedel-de Haën, Seelze, Germany) at 1% of acetic acid (Merck, Darmstadt, Germany) and 10% methanol (Scharlau, Sentmenat, Spain); flow rate, 0.7 mL/min; injection volume, 20 µL [26]. The wavelength range was 220–660 nm and the chromatograms were monitored at 285 nm. HMF and 2F were identified by comparison of the UV spectra and the retention time with those of HMF and 2F standard ($p \geq 99$ % Sigma-Aldrich, St. Louis, MO, USA) and quantified using an external calibration curve. All analyses were performed in duplicate, including the sample dilution procedure, and the reported HMF and 2F concentration is therefore the average of four values.

Kinetic parameters for HMF and 2F formation in MdL stored under different lighting conditions and at different temperatures were calculated as reported by Arena et al. [25]. The activation energies Ea (kcal mol^{-1}) values of HMF and 2F were calculated from rate coefficients at different temperatures by applying the Arrhenius equation.

2.5. *Total Phenols*

The total phenols (TP) analysis was performed as reported by Di Stefano et al. [38].

Briefly, 10 mL of wine was diluted 1:2 with 1 N H_2SO_4. Then, the diluted sample was passed through Sep Pack C18 cartridges (Waters Chromatography Europe BV, Etten-Leur, The Netherlands) previously activated with methanol and distilled water (2 and 5 mL, respectively). Adsorbed phenols were washed with 2 mL of 0.1 N H_2SO_4 and then desorbed with methanol (3 mL). The TP were assessed by the colorimetric Folin–Ciocalteu method at 700 and 760 nm, with gallic acid as calibration standard. Results were expressed as mg L^{-1}.

2.6. Chemicals and Reagents

All reagents and solvents HPLC grade were purchased from Merck (Darmstadt, Germany). HMF, 2F and gallic acid standards were from Sigma (Milano, Italy).

2.7. Statistical Analysis

Statistical analyses were performed to assess the effect of packaging and storage conditions on the analytical parameters. The significance of differences was estimated by analysis of variance (ANOVA). The statistical significance level was set to 0.05. Statistical analyses were performed using SPSS® Statistics 13.0 (IBM, Armonk, NY, USA).

Results are presented as mean value ± standard deviation. Data were analyzed through Spearman's correlation to have a measure of the strength and direction of the association or relationship between the concentrations of analytical parameters and time of storage under different CWF lamp, and time of storage under different temperatures. All pairwise comparisons were run at 95% confidence intervals and p-values were Bonferroni adjusted through the statistical package SPSS® Statistics 13.0 (IBM, Armonk, NY, USA).

3. Results and Discussion

3.1. Light Transmission Properties of Clear Glass Bottles

Figure 1 shows the light shielding expressed as percentage. As expected, the clear, colorless bottles offered a lower shielding effect in all spectra regions considered. In particular, shielding increased in the following order: colorless (10%), green (26.7%) and amber (49%) bottles for the illuminance parameter. A similar trend was observed for the UVA irradiance, while the shielding effect in the UVB region was 100% for both colored bottles and close to 90% for the clear bottle. This result is in agreement with results of Maury et al. [17], who compared the transmission spectrum of clear and green bottle colors and found that the former was capable of transmitting all visible and some UV light.

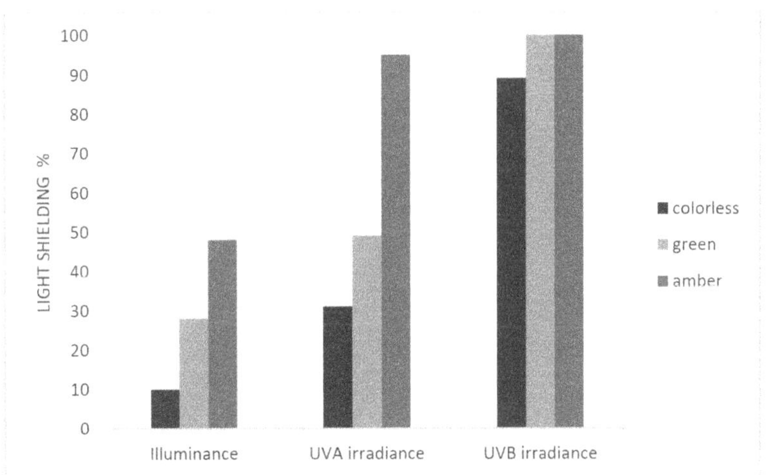

Figure 1. Light shielding (%) of clear glass bottles (colorless, green, amber).

3.2. Effect of Light Exposure and Bottle Color on the Quality of Malvasia Delle Lipari Wine

MdL was characterized by the following base analytical parameters: pH = 3.79; °Brix = 1.35; total acidity = 5.4 g L^{-1}; volatile acidity = 0.95 g L^{-1}; alcohol content = 15.8% (v/v); TP = 247.10 mg L^{-1}; HMF = 1.16 mg L^{-1}; 2F = 1.11 mg L^{-1}; L^* = 94.90; a^* = −2.99; b^* = 17.63; C = 17.88; h = 0.17.

Table 3 reports the changes in the quality parameters, as influenced by light exposure and glass bottle color over 90 days of storage.

Table 3. Quality parameters of MdL wine samples as influenced by light exposure (number of CWF lamps) and clear-colored glass bottles (A = amber; G = green; C = colorless) during 90 days' storage (at constant temperature T = 25 ± 2 °C). (TP: total phenols; C: chroma; h: hue; L^*: lightness; a^*: redness; b^*: yellowness; HMF: 5-hydroxymethylfurfural; 2F: 2-furaldehyde).

Time (Days)	CWF Lamp	Bottle Color	TP (mg L^{-1})	C	h	L^*	a^*	b^*	HMF (mg L^{-1})	2F (mg L^{-1})
30	1	A	163.90 ± 3.40 ^{mn}	18.76 ±0.3 ^{de}	−0.11 ± 0.02 ^{c}	94.22 ± 0.44 ^{bc}	−2.14 ± 0.06 ^{b}	18.63 ± 0.07 ^{de}	2.08 ± 0.07 ^{g}	1.02 ± 0.04 ^{f}
		G	168.62 ± 0.88 ^{l}	18.06 ± 1.1 ^{e}	−0.09 ± 0.01 ^{cd}	93.78 ± 0.55 ^{bcd}	−1.62 ± 0.00 ^{a}	17.99 ± 0.15 ^{e}	2.23 ± 0.07 ^{g}	1.24 ± 0.08 ^{ef}
		C	158.31 ± 0.46 ^{o}	17.41 ± 0.4 ^{e}	−0.12 ± 0.01 ^{bc}	94.60 ± 0.69 ^{bc}	−2.01 ± 0.04 ^{b}	17.29 ± 0.38 ^{e}	1.89 ± 0.08 ^{gh}	1.32 ± 0.09 ^{e}
	4	A	179.40 ± 2.25 ^{i}	19.97 ± 0.4 ^{d}	−0.12 ± 0.2 ^{bc}	93.69 ± 0.92 ^{bcd}	−2.38 ± 0.08 ^{bc}	19.82 ± 1.18 ^{d}	4.62 ± 0.05 ^{ef}	1.57 ± 0.09 ^{d}
		G	160.94 ± 1.33 ^{m}	19.06 ± 0.0 ^{d}	−0.11 ± 0.01 ^{c}	95.04 ± 2.56 ^{b}	−2.00 ± 0.20 ^{b}	18.96 ± 0.68 ^{de}	3.51 ± 0.08 ^{fg}	1.90 ± 0.09 ^{c}
		C	144.06 ± 2.01 ^{p}	18.79 ± 1.4 ^{de}	−0.16 ± 0.3 ^{ab}	94.79 ± 1.32 ^{bc}	−3.00 ± 0.42 ^{bc}	18.55 ± 1.66 ^{de}	3.75 ± 0.07 ^{fg}	1.50 ± 0.07 ^{ed}
	6	A	193.26 ± 4.21 ^{gh}	19.81 ± 2.0 ^{d}	−0.11 ± 0.01 ^{c}	93.23 ± 0.34 ^{bcd}	−2.07 ± 0.01 ^{b}	19.70 ± 0.49 ^{d}	4.23 ± 0.05 ^{f}	1.85 ± 0.09 ^{c}
		G	190.31 ± 1.56 ^{g}	19.50 ± 1.0 ^{d}	−0.15 ± 0.1 ^{b}	98.64 ± 6.29 ^{a}	−2.85 ± 0.42 ^{bc}	19.29 ± 0.36 ^{d}	4.55 ± 0.08 ^{f}	1.61 ± 0.07 ^{d}
		C	192.51 ± 2.06 ^{gh}	18.13 ± 0.2 ^{de}	−0.16 ± 0.1 ^{ab}	94.73 ± 0.51 ^{bc}	−2.95 ± 0.39 ^{bc}	17.89 ± 0.54 ^{e}	3.98 ± 0.07 ^{efg}	1.34 ± 0.09 ^{e}
60	1	A	190.58 ± 0.94 ^{g}	22.33 ± 0.1 ^{c}	−0.17 ± 0.2 ^{ab}	94.31 ± 0.02 ^{bc}	−3.75 ± 0.01 ^{cd}	22.01 ± 0.00 ^{c}	4.26 ± 0.47 ^{f}	1.06 ± 0.01 ^{f}
		G	189.52 ± 1.10 ^{g}	21.74 ± 0.1 ^{cd}	−0.16 ± 0.01 ^{ab}	94.23 ± 0.00 ^{bc}	−3.41 ±0.00 ^{c}	21.47 ± 0.01 ^{cd}	4.28 ± 0.20 ^{f}	1.62 ± 0.33 ^{d}
		C	225.34 ± 2.95 ^{c}	21.02 ± 2.3 ^{cd}	−0.18 ± 0.0.14 ^{a}	94.79 ± 0.01 ^{bc}	−3.70 ± 0.01 ^{cd}	20.70 ± 0.02 ^{cd}	3.68 ± 0.12 ^{fg}	1.70 ± 0.22 ^{cd}
	4	A	205.22 ± 4.21 ^{f}	24.83 ± 0.4 ^{bc}	−0.20 ± 0.1 ^{a}	95.29 ± 0.14 ^{b}	−4.90 ± 0.02 ^{de}	24.34 ± 0.01 ^{bc}	18.56 ± 0.37 ^{c}	2.03 ± 0.26 ^{bc}
		G	210.42 ± 3.05 ^{e}	24.61 ± 2.1 ^{bc}	−0.18 ± 0.1 ^{a}	94.54 ± 0.10 ^{bc}	−4.25 ± 0.01 ^{d}	24.24 ± 0.00 ^{bc}	12.65 ± 2.24 ^{cd}	2.65 ± 0.04 ^{bc}
		C	220.26 ± 0.45 ^{d}	22.83 ± 0.1 ^{c}	−0.19 ± 0.02 ^{a}	95.18 ± 0.10 ^{b}	−4.31 ± 0.31 ^{d}	22.42 ± 0.30 ^{c}	14.01 ± 0.93 ^{cd}	2.23 ± 0.04 ^{bc}
	6	A	210.22 ± 2.01 ^{e}	24.03 ± 0.4 ^{bc}	−0.18 ± 0.1 ^{a}	94.62 ± 0.01 ^{bc}	−4.19 ± 0.00 ^{d}	23.32 ± 0.02 ^{bc}	11.37 ± 1.04 ^{cd}	2.92 ± 0.53 ^{ab}
		G	219.75 ± 1.32 ^{de}	22.98 ± 0.2 ^{c}	−0.18 ± 0.1 ^{a}	94.70 ± 0.00 ^{bc}	−4.15 ± 0.01 ^{d}	22.70 ± 0.00 ^{c}	10.44 ± 0.71 ^{d}	2.32 ± 0.10 ^{b}
		C	197.33 ± 0.47 ^{fg}	25.25 ± 1.4 ^{bc}	−0.19 ± 0.1 ^{a}	95.11 ± 0.11 ^{b}	−4.34 ± 0.00 ^{d}	23.56 ± 0.01 ^{bc}	10.80 ± 0.05 ^{d}	1.83 ± 0.04 ^{c}
90	1	A	236.10 ± 0.70 ^{b}	29.59 ± 0.4 ^{ab}	−0.11 ± 0.1 ^{c}	91.89 ± 0.21 ^{d}	−3.20 ± 0.23 ^{cd}	29.42 ± 0.11 ^{ab}	7.81 ± 1.21 ^{e}	1.70 ± 0.09 ^{cd}
		G	216.32 ± 1.74 ^{de}	26.28 ± 0.1 ^{b}	−0.15 ± 0.1 ^{b}	93.49 ± 0.20 ^{bcd}	−3.96 ± 0.12 ^{cd}	25.98 ± 0.20 ^{b}	9.92 ± 0.87 ^{de}	1.81 ± 0.02 ^{c}
		C	297.54 ± 0.40 ^{a}	26.64 ± 0.2 ^{b}	−0.14 ± 0.1 ^{bc}	93.35 ± 0.20 ^{bcd}	−3.76 ± 0.14 ^{cd}	26.37 ± 0.41 ^{b}	8.35 ± 0.47 ^{de}	1.68 ± 0.18 ^{cd}
	4	A	227.30 ± 3.74 ^{c}	32.05 ± 0.4 ^{a}	−0.16 ± 0.1 ^{ab}	93.08 ±0.22 ^{bcd}	−4.95 ±0.11 ^{de}	31.66 ± 0.30 ^{ab}	26.82 ± 0.87 ^{b}	3.26 ± 0.05 ^{a}
		G	228.84 ± 2.31 ^{c}	29.94 ± 0.5 ^{ab}	−0.15 ± 0.01 ^{b}	93.01 ± 0.30 ^{bcd}	−4.31 ± 0.10 ^{d}	29.62 ± 0.21 ^{ab}	27.35 ± 0.38 ^{b}	2.97 ± 0.21 ^{ab}
		C	236.14 ± 1.08 ^{b}	35.94 ± 1.4 ^{a}	−0.04 ± 0.01 ^{d}	88.12 ± 0.43 ^{de}	−1.60 ±0.00 ^{a}	35.90 ± 0.20 ^{a}	26.58 ± 0.85 ^{b}	3.58 ± 0.15 ^{a}
	6	A	224.31 ± 0.85 ^{cd}	30.24 ± 2.1 ^{ab}	−0.16 ± 0.1 ^{ab}	93.39 ± 0.51 ^{bcd}	−4.84 ± 0.31 ^{de}	29.85 ± 1.10 ^{ab}	36.43 ± 0.97 ^{a}	3.00 ± 0.10 ^{ab}
		G	220.57 ± 2.45 ^{d}	28.03 ± 0.2 ^{ab}	−0.15 ± 0.1 ^{b}	92.78 ± 0.20 ^{d}	−4.26 ± 0.30 ^{d}	27.70 ± 1.00 ^{b}	33.74 ± 0.10 ^{ab}	2.82 ± 0.06 ^{b}
		C	204.18 ± 1.04 ^{f}	25.39 ± 0.4 ^{bc}	−0.17 ± 0.1 ^{ab}	94.26 ± 0.90 ^{bc}	−4.35 ± 0.33 ^{d}	25.01 ± 1.5 ^{b}	30.75 ± 0.90 ^{ab}	3.04 ± 0.18 ^{ab}

Different letters within the same parameter and main factor show (bottle's color) significant differences (LSD test, $p \leq 0.05$).

Comparing the values of volatile acidity and pH at time "0", both parameters showed the same course in all clear-colored glass bottles under all lighting conditions, with a slight decrease for volatile acidity up to 0.82 ± 0.02 g L^{-1}. Total acidity ranged between 5.40 ± 0.34 g L^{-1} and 5.23 ± 0.06 g L^{-1} in all clear-colored glass bottles; samples bottled in clear colorless glass showed the same trend under 1, 4 and 6 CWF lamp lighting. Overall, small variations were observed during the shelf-life tests for pH, volatile and total acidity, but no statistical significance was observed ($p > 0.05$). Similarly, Revi et al. [18] as well as Hopfer et al. [39] found no differences ($p > 0.05$) in total and volatile acidity, pH and ethanol content of Chardonnay wine between bag-in-box containers and glass screw cap bottles after three months of storage at 20 °C.

Arapitsas et al. [40], studying white wine light-strike fault in flint and green glass bottles, reported any statistically significant difference among pH, titratable acidity and volatile acidity, and as the last one was not affected by time or packaging choice.

The trend for the TP during storage was the same for all clear-colored glass bottles and lighting conditions tested, reaching a mean value of 249.11 ± 18.66 mg L^{-1} after 90 days.

These results are discordant with previous observations where we noticed a reduction in TP during wine ageing, explained by the transformation and/or precipitation of phenolic material as oxidation reactions progress [41].

In MdL, this parameter did not change significantly, neither with different light exposure, nor with storage time ($p > 0.05$).

Color is one of the major attributes that affects the consumer perception of quality. Small changes in CIE parameters were observed. In particular, C, hue and b^* progressively increased at each lighting condition, from 0 to 90 days of the study. For instance, C changed during the storage time by a percentage of 11, 39 and 79% after 30, 60 and 90 days, respectively, in amber bottles under 4 CWF lamps. Similarly, the change ranged to about 7, 37 and 67% after 30, 60 and 90 days, respectively, in green bottles under 4 CWF lamps and up to 100% for colorless bottles under the same lighting. The b^* value increased proportionally through the storage period with the increasing number of CWF lamps, with R2 0.87, 0.90 and 0.95 after 30, 60 and 90 days, respectively. The increase in b^* values represents a higher intensity of yellow and the main increase is always in amber bottles, with only the exception of colorless bottles observed after 90 days and exposed under 4 CWF lamps. The L^* values, which give a value of the brightness of the samples, slightly decrease during 90 days of storage, until the 2–3% of reduction from the initial value for clear green and clear amber bottles under 6 and 1 CWF lamps respectively, while the decrease is near the 8% in colorless bottles exposed at 4 CWF lamps. The a^* value, which represents the degree of one component of green color, had a reduction during the first 30 days, and then an increase was observed after 60 days of storage under illumination, especially in clear amber bottles stored under 4 CWF lamps, according to Refsgaard et al. [42], while MdL stored in clear green bottles did not show any particular differences from the beginning. Hue (h), the attribute of appearance by which a color is identified according to its resemblance to red, green, yellow or blue, slightly decreases from 30 to 90 days. An increase in the yellow color (b^*) and hue is usually associated with wine aging, as well as a decrease in a^*, measured in light-exposed samples [43].

The HMF content showed a similar trend for the clear amber, green and colorless bottles, while some differences were observed when MdL was stored under different lighting conditions, mostly at 4 and 6 CWF lamps. In these cases, a rapid HMF increase was observed in the first 60 days, up to 18.56 ± 0.37 mg L^{-1}, with a progressive increase over the following 30 days, up to 36.43 ± 0.97 mg L^{-1} (Table 3). Based on two-way ANOVA, the interaction effect of bottle color × lighting conditions on HMF concentration was not significant, while a significant effect ($p < 0.001$) was found for the lighting conditions, as expressed with different letters in Figure 2. It is well known that several factors influence the formation rate of this compound, such as temperature, time and storage conditions [26], as well as sugar concentration and water activity [44]. HMF reacted with ethanol to form 5-(ethoxymethyl) furfural, a compound founded in sweet fortified white wine, with a concentration above of the perception threshold, after 6 months of aging, in the absence of air [45].

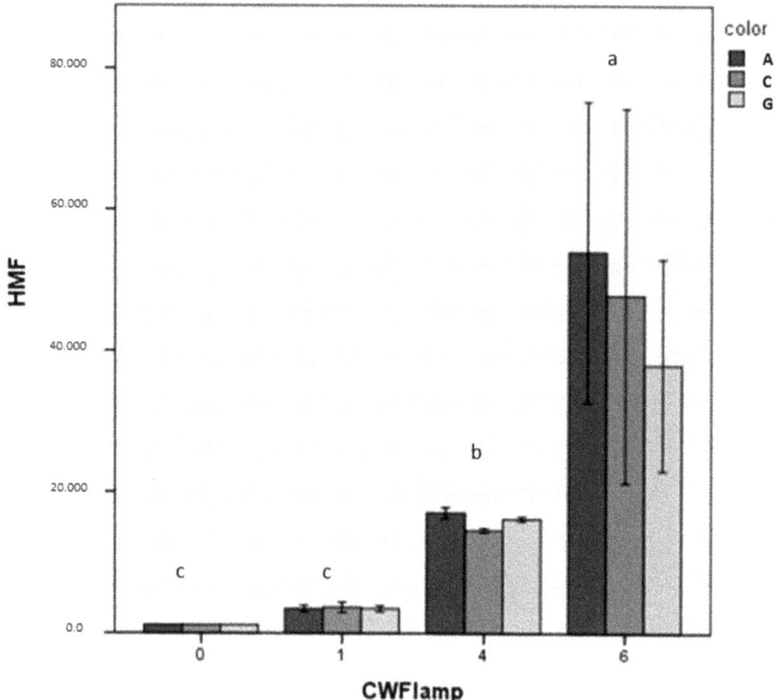

Figure 2. HMF concentration at constant temperature of 25 °C influenced by the number of CWF lamps (0, 1, 4, 6) and the color of clear glass bottles (amber, colorless, green); different letters represent significant differences (LSD test, $p \leq 0.05$).

As concerns 2F, produced by breakdown of pentose and/or Maillard reaction, from the starting value of 2F (1.11 mg L^{-1}), an increase in the 2F concentration was observed in samples kept under 4 and 6 CWF lamps, until the final values were 3.3 $\pm$ 0.3 mg L^{-1} and 2.9 $\pm$ 0.2 mg L^{-1}, respectively. Similarly to HMF, a statistically significant effect of lighting on 2F concentration was observed F (3.30 = 5.473, p = 0.007, partial η^2 =0.477).

A two-way ANOVA was conducted to examine the effects of the exposure to a different number of CWF lamps, and the color of the glass bottles on each dependent variable examined (total and volatile acidity, TP, HMF, 2F, color parameter and pH). Residual analysis was performed to test for the assumptions of the two-way ANOVA. Outliers were assessed by inspection of a boxplot, then data were verified for the non-normality and homogeneity of variances. Normality was assessed using Shapiro–Wilk's normality test for each cell of the design, and the homogeneity of variances was assessed by Levene's test.

The interaction effect between CWF lamps and bottle color (Table 3) on volatile acidity, TP, Chroma, Hue and pH, was not statistically significant (F-test—degrees of freedom for the interaction term 6 in (6, 69)). Finding a non-statistically significant interaction does not mean that an interaction effect does not exist. Therefore, an analysis of the main effect for a kind of lighting (number of CWF lamp) was performed, which indicated that the main effect was not statistically significant (p > 0.05). All pairwise comparisons run where reported 95% confidence intervals and p-values are Bonferroni-adjusted.

At constant temperature, a Spearman's rank-order correlation (2-tailed) was run to assess the relationship between HMF and 2F concentration, with storage time under 1,

4 and 6 CWF lamp. Preliminary analysis showed the relationship to be monotonic, as assessed by the visual inspection of a scatterplot. There was a strong positive correlation between HMF and the number of CWF lamps, $r_s = 0.846$, $p < 0.01$ (Figure 2); as well as a positive correlation between 2F and storage time, $r_s = 0.429$, $p < 0.05$.

Table 4 reports the kinetics parameters of both HMF and 2F formation in MdL wine stored under different lighting conditions. All systems fit a pseudo-first-order equation [46] well, and *k* values for HMF formation were always higher respect to those determined for 2F formation, regardless of bottle color and lighting conditions.

The MdL samples stored under 1 CWF lamp showed the lowest *k* values for HMF formation (about 0.017 days^{-1}) independently from the bottle color. As the light exposure increases from 1 to 4 CWF lamps, the *k* value for HMF formation increases up to about 0.034 days^{-1} and was similar to those determined for MdL stored under 6 CWF lamps (about 0.031 days^{-1}).

Similarly to HMF, 2F shows the lowest *k* value under 1 CWF lamp, while it increased by 2.6 fold under the lightening of 4 and 6 CWF lamp, independently of the bottle color.

Table 4. Kinetics parameters for HMF and 2-furaldehyde formation in MdL stored in clear-colored glass bottles (amber, green, colorless) under different lighting condition (at constant temperature T = 25 ± 2 °C). (Control was a sample kept in the dark).

	5-Hydroxymethylfurfural					
	1 CWF Lamp		**4 CWF Lamps**		**6 CWF Lamps**	
	***k* (min)**	**R^2**	***k* (min)**	**R^2**	***k* (min)**	**R^2**
Control	0.0116	0.91296	0.0116	0.91296	0.0116	0.91296
Amber	0.0164	0.95525	0.0363	0.98405	0.0316	0.99903
Green	0.0183	0.95012	0.0333	0.98975	0.0315	0.99805
Colorless	0.016	0.90128	0.0333	0.98967	0.0312	0.99953
	2-Furaldehyde					
	1 CWF Lamp		**4 CWF Lamps**		**6 CWF Lamps**	
	***k* (min)**	**R^2**	***k* (min)**	**R^2**	***k* (min)**	**R^2**
Control	0.0044	0.7529	0.0044	0.7529	0.0044	0.7529
Amber	0.0048	0.9999	0.0115	0.9855	0.013	0.8765
Green	0.0056	0.967	0.0125	0.8969	0.0111	0.9769
Colorless	0.0054	0.8776	0.0125	0.9878	0.0101	0.9429

3.3. *Effect of Storage Temperature on Quality Changes of Malvasia Delle Lipari Wine*

After the first 90 days, samples were stored up to 180 days under constant illumination (one CWF lamp) at three different temperatures: 30, 35 and 40 °C (Table 5).

The pH value was stable during the first 90 days of the study, showing a slight decrease, reaching 180 days, especially at 30 °C and 35 °C, reaching values of 3.31 and 3.29 respectively; in addition, volatile acidity showed only slight changes and limited standard deviation.

The TP content at the end of the study was practically coincident with the initial values, notwithstanding the oscillations observed during the trials.

The Chroma (C) is the measure of a stimulus judged relative to the white. Chroma, hue and b^* followed the same path. Initial C value was 35.10 and its value increased from 90 to 180 days until an average value of 46.00 ± 8.30 among the three temperatures. C values increased with storage temperature, in the order: 30, 35 and 40 °C. No significant change was observed for L^*. The a^* value was stable for the first 150 days to increase in samples at 35 °C and 40 °C. From 150 to 180 days, an increase was also monitored in MdL stored at 30 °C, exceeding the values of all the other samples. As previously reported, an increase in color parameters is linked to light-exposed samples [43].

Table 5. Evolution of quality parameters of MdL wine samples at different storage temperatures (30, 35, 40 °C) (HMF: 5-hydroxymethylfurfural; 2F: 2-furaldehyde; TP: total phenols; C: chroma; h: hue; L^*: lightness) from 90 to 180 days.

Time (Days)	T °C	HMF mg L^{-1}	2F mg L^{-1}	TP mg L^{-1}	C	h	L^*
90	30	8.69 ± 1.10 ^{h}	1.73 ± 0.07 ^{e}	259.12 ± 1.62 ^{bc}	31.95 ± 0.70 ^{g}	−0.14 ± 0.00 ^{a}	92.15 ± 0.27 ^{a}
	35	26.92 ± 0.40 ^{fgh}	3.27 ± 0.30 ^{cde}	263.92 ± 9.23 ^{bc}	37.61 ± 2.97 ^{de}	−0.12 ± 0.00 ^{bc}	91.13 ± 0.49 ^{a}
	40	33.64 ± 2.84 ^{fg}	2.95 ± 0.12 ^{de}	250.02 ± 9.06 ^{c}	35.75 ± 3.46 ^{e}	−0.12 ±0.00 ^{bc}	90.88 ± 0.86 ^{ab}
120	30	17.60 ± 0.56 ^{gh}	2.10 ±0.54 ^{e}	257.62 ± 26.1 ^{bc}	33.19 ± 0.15 ^{ef}	−0.13 ± 0.01 ^{ab}	91.98 ± 0.49 ^{a}
	35	55.86 ± 2.19 ^{de}	3.71 ± 0.33 ^{cde}	249.62 ± 10.6 ^{cd}	41.82 ± 0.41 ^{c}	−0.10 ± 0.01 ^{cd}	90.00 ± 0.39 ^{ab}
	40	68.23 ±3.50 ^{d}	4.71 ± 0.31 ^{bcd}	284.77 ± 0.90 ^{a}	38.47 ± 0.31 ^{d}	−0.11 ± 0.01 ^{c}	90.13 ± 1.33 ^{ab}
150	30	29.70 ± 1.18 ^{fg}	1.88 ± 0.19 ^{e}	231.00 ± 4.53 ^{ef}	34.62 ± 1.11 ^{ef}	−0.13 ± 0.02 ^{ab}	91.64 ± 0.48 ^{a}
	35	101.43 ± 3.0 ^{c}	5.24 ± 0.45 ^{abc}	234.76 ± 30.48 ^{e}	50.58 ± 1.94 ^{ab}	−0.06 ± 0.02 ^{e}	87.89 ± 1.13 ^{bc}
	40	107.25 ± 9.7 ^{bc}	5.07 ± 0.32 ^{abcd}	246.72 ± 3.44 ^{cd}	41.04 ± 4.32 ^{c}	−0.09 ± 0.00 ^{d}	89.86 ± 0.28 ^{b}
180	30	37.13 ± 1.86 ^{ef}	2.02 ± 0.31 ^{e}	238.42 ± 10.37 ^{d}	40.76 ± 7.42 ^{cd}	−0.06 ± 0.11 ^{e}	87.24 ± 8.10 ^{bc}
	35	124.49 ± 15.54 ^{b}	7.17 ± 1.76 ^{a}	269.92 ± 16.60 ^{ab}	55.57 ± 5.52 ^{a}	0.05 ± 0.12 ^{e}	76.26 ± 15.5 ^{d}
	40	152.81 ± 12.18 ^{a}	6.32 ± 1.69 ^{ab}	239.00 ± 6.00 ^{d}	41.67 ± 4.92 ^{c}	−0.10 ± 0.02 ^{cd}	89.60 ± 1.34 ^{b}

Different letters within the same parameter and main factor show significant differences (LSD test, $p \leq 0.05$).

The HMF levels changed considerably, mostly when MdL was stored at the highest temperature. At 30 °C, the HMF content increased from about 9 mg/kg up to 37 mg/kg at 180 days of storage. At 35 °C, the HMF content increased up to about 125 mg/kg at the end of storage. The effect of temperature on HMF formation, obviously, was more evident at 40 °C. At this storage, the HMF increases rapidly, even after 120 days, and at 180 days of storage, reached the highest levels, about 153 mg/kg.

During MdL storage at 30 and 35 °C, 2F increased slowly and important changes were only founded at 40 °C. At this temperature, after 180 days of storage, the 2F concentration was about 3 times higher than that determined at 30 °C.

Moreover, HMF levels were always higher than 2F. The first furanic compound was derived from glucose and fructose, while 2F comes from pentoses present in wine, in minor amounts. This trend was reported by other authors in sweet wine [30].

To underline changes related to different storage temperatures, a Spearman's rank-order correlation was carried out to assess the relationship between color parameters, HMF and 2F concentration, with storage time under different temperatures. Preliminary analysis showed the relationship to be monotonic, as assessed by visual inspection of a scatterplot. There were strong positive correlations ($p < 0.01$) between: HMF and 2F, $r_s = 0.972$; HMF and temperatures, $r_s = 0.798$; 2F and temperatures, $r_s = 0.739$ as reported in Table 6 (the correlation between C and all the color parameters was significant for $p < 0.01$).

Table 7 shows the kinetic parameters k (days^{-1}), and the activation energy (Ea), both for HMF and 2F formation, in MdL wine stored at different temperatures. Concerning HMF, the lowest k value was determined at 30 °C (0.0144 days^{-1}), and the highest was found at 40 °C (0.0167 days^{-1}).

The kinetics for 2F formation exhibited a k value lower than those of HMF, ranging from 0.0038 to 0.0088 days^{-1}, at 30 and 40 °C, respectively.

Cutzach et al. [45] studied the formation of some compounds, including HMF and 2F, during the aging of sweet fortified wines at the temperature of 37 °C. Data reported in this work concerning HMF and 2F were used to determine k value, in order to compare our data with those reported in literature. Furthermore, in this case, k value was lowest for 2F formation and highest for HMF formation and kinetic parameters were very similar to our data (2F, $k = 0.0099$ days^{-1}, $R^2 = 0.987$; HMF, $k = 0.0136$ days $^{-1}$, $R^2 = 0.992$), suggesting a similar mechanisms of reactions.

As concerns the activation energy (Table 7), the lowest value was determined for HMF (11.7 kJ mol^{-1}), and the highest, about 5.7-fold higher than HMF, for 2F formation

(66.4 kJ mol^{-1}), suggesting that a small increase of temperature induced a fast increase of the formation of these compounds, mostly for HMF.

Table 6. Spearman's rho T = 30-35-40 °C from 90 to 180 days of storage (N = 12). (HMF: 5-hydroxymethylfurfural; 2F: 2-furaldehyde: C: chroma; h: hue; L^*: lightness; a^*: redness; b^*: yellowness).

		HMF	2-Furaldehyde	C	h	L^*	a^*	b^*
Time (days)	Correlation coefficient	0.518 *	0.497	0.626 *	0.718 **	−0.777 *	0.791 **	0.626 *
	Sig. (2-tailed)	0.084	0.101	0.029	0.009	0.003	0.002	0.029
T °C	Correlation coefficient	0.798 **	0.739 **	0.414	0.313	−0.0325	0.312	0.414
	Sig. (2-tailed)	0.002	0.006	0.181	0.322	0.302	0.324	0.181
HMF	Correlation coefficient		0.972 **					
	Sig. (2-tailed)		0.000					
C	Correlation coefficient				0.919 **	−0.902 **	0.804 **	1.000 **
	Sig. (2-tailed)			.	0.000	0.000	0.002	
h	Correlation coefficient					−0.982 **	0.942 **	0.919 **
	Sig. (2-tailed)					0.000	0.000	0.000
L^*	Correlation coefficient						−0.951 **	−0.902 **
	Sig. (2-tailed)						0.000	0.000
a^*	Correlation coefficient							−0.804 **
	Sig. (2-tailed)							0.002

** Correlation is significant at the 0.01 level (2-tailed); * Correlation is significant at the 0.05 level (2-tailed).

Table 7. Kinetic parameters for 5-hydroxymethylfurfural (HMF) and 2-furaldehyde (2F) formation in MdL stored under different temperatures (30, 35, 40 °C).

HMF				2F			
T (°C)	***k* days^{-1}**	**R^2**	**Ea kJ mol^{-1} (kcal mol^{-1})**	**T (°C)**	***k* days^{-1}**	**R^2**	**Ea kJ mol^{-1} (kcal mol^{-1})**
30 °C	0.0144	0.9538		30 °C	0.0038	0.6275	
35 °C	0.0150	0.9448	11.7 (2.8)	35 °C	0.0070	0.9010	66.4 (15.9)
40 °C	0.0167	0.9872		40 °C	0.0088	0.9488	

In recent years, much research has been focused on the influence of light exposure to wine composition, in particular, on white wine, but also on red wine [41,43]. The phenomenon of light-induced off-flavors in wine, called "light-struck taste" (LST) [40,46], together with bottle color, storage condition and temperature effect were studied to improve the quality maintenance of wine during its storage [40,43,47–49]. To the best of our knowledge, no one has studied light interactions and bottle's color, or the temperature's effect on HMF and 2F, during a prolonged storage of 90 and 180 days, respectively.

4. Conclusions

The study was conducted in order to assess the changes in some quality attributes of MdL wine as a function of clear-colored glass bottles, lighting and temperature. Overall, the quality parameters of MdL wine were not significantly altered with the exposure to different lighting conditions and under different storage conditions, however, some changes in the color parameters occurred in every tested condition. In contrast with common beliefs, the bottle color did not play a significant role in the quality maintenance of this sweet wine. The significant increase in the levels of HMF and 2F was mainly dependent on the intensity of lighting and on the storage temperature, as confirmed by the activation energy, without noticeable effects of the bottle color. Results suggest that retailers have a higher responsibility in the product quality maintenance compared to the company's choices of bottle color, which, in the specific case of MdL wine, can be merely driven by marketing consideration. Finally, HMF and 2F levels can be considered as global quality indicators for MdL wine during storage, offering a tool for detecting unsuitable storage conditions.

In conclusion, the quality maintenance of wine can be achieved by controlling environmental factors, such as light exposure and temperature, during its storage guaranteeing profits for wine producers and satisfaction for consumers.

Author Contributions: Conceptualization: G.M. F.L. B.F.; Formal analysis: E.A.; V.R.; F.L.; Writing—original draft: V.R.; F.L.; Writing—review and editing: E.A.; V.R.; F.L.; B.F; G.M. All authors have read and agreed to the published version of the manuscript.

Funding: The research was supported by the Project "Valutazione della sostenibilità dei sistemi agroalimentari locali e selezione di markers molecolari e biologici nella gestione della qualità di prodotti agroalimentari" funded by the University of Catania (Italy).

Acknowledgments: The authors thank Carlo Nicolosi Asmundo and Marco Nicolosi for kindly supplying Malvasia delle Lipari DOC wine.

Conflicts of Interest: The authors declare no conflict of interest.

References

1. Del Nobile, M.; Ambrosino, M.; Sacchi, R.; Masi, P. Design of Plastic Bottles for Packaging of Virgin Olive Oil. *J. Food Sci.* **2003**, *68*, 170–175. [CrossRef]
2. Piergiovanni, L.; Limbo, S. The protective effect of film metallization against oxidative deterioration and discoloration of sensitive foods. *Packag. Technol. Sci.* **2004**, *17*, 155–164. [CrossRef]
3. Licciardello, F.; Del Nobile, M.A.; Spagna, G.; Muratore, G. Scalping of ethyloctanoate and linalool from a model wine into plastic films. *LWT* **2009**, *42*, 1065–1069. [CrossRef]
4. Aldave, L.; Almy, J.; Cabezudo, M.D.; Caceres, I.; Gonzales-Raurich, M.; Salvador, M.D. The shelf life of young white wines. In *Shelf Life Studies of Foods and Beverages*; Charalambous, G., Ed.; Elsevier: London, UK, 1993; pp. 923–943.
5. Sanova-Savova, S.; Dimov, S.; Ribarova, F. Anthocyanins and Color Variables of Bulgarian Aged Red Wines. *J. Food Compos. Anal.* **2002**, *15*, 647–654. [CrossRef]
6. Singleton, V.L. Oxygen with phenols and related reactions in musts, wines and model systems: Observations and practical implications. *Am. J. Enol. Vitic.* **1987**, *38*, 69–76.
7. Cheynier, V.; Rigaud, J.; Souquet, J.M.; Barillére, J.M.; Moutounet, M. Effect of pomace contact and hyperoxidation on the phenolic composition and quality of Grenache and Chardonnay wines. *Am. J. Enol. Vitic.* **1989**, *40*, 36–42.
8. Cheynier, V.; Rigaud, J.; Souquet, J.M.; Barillére, J.M.; Moutounet, M. Must browning in relation to the behaviour of phenolic compounds during oxidation. *Am. J. Enol. Vitic.* **1990**, *41*, 346–349.
9. Sims, C.; Morris, J. The effect of pH, sulfur dioxide, storage time and temperature on the color and stability of red muscadine grape wine. *Am. J. Enol. Vitic.* **1984**, *35*, 35–39.
10. Gomez-Plaza, E.; Minano, A.; Lopez-Roca, J.M. Comparison of chromatic properties, stability and antioxidant capacity of anthocyanin-based aqueous extracts from grape pomace obtained from different vinification methods. *Food Chem.* **2006**, *97*, 87–94. [CrossRef]
11. Marquez, A.; Serratosa, M.P.; Merida, J. Influence of bottle storage time on colour, phenolic composition and sensory properties of sweet red wines. *Food Chem.* **2014**, *146*, 507–514. [CrossRef] [PubMed]
12. Recamales, A.F.; Sarago, A.; Gonzalez-Miret, M.L.; Hernanz, D. The effect of time and storage conditions and colour of white wine. *Food Res. Int.* **2006**, *39*, 220–229. [CrossRef]
13. Kallithraka, S.; Salacha, M.I.; Tzourou, I. Changes in phenolic composition and antioxidant activity of white wine during bottle storage: Accelerated browning test versus bottle storage. *Food Chem.* **2009**, *113*, 500–505. [CrossRef]
14. Selli, S.; Canbas, A.; Unal, U. Effect of bottle colour and storage conditions on browning of orange wine. *Nahrung/Food* **2002**, *46*, 64–67. [CrossRef]
15. Benitez, P.; Castro, R.; Garcia Barroso, C. Changes in the Polyphenolic and Volatile Contents of "Fino" Sherry Wine Exposed to Ultraviolet and Visible Radiation during Storage. *J. Agric. Food Chem.* **2003**, *51*, 6482–6487. [CrossRef]
16. Dias, D.A.; Smith, T.A.; Ghiggino, K.P.; Scollary, G.R. The role of light, temperature and wine bottle colour on pigment enhancement in white wine. *Food Chem.* **2012**, *135*, 2934–2941. [CrossRef]
17. Maury, C.; Clark, A.C.; Scollary, G.R. Determination of the impact of bottle colour and phenolic concentration on pigment development in white wine stored under external conditions. *Anal. Chim. Acta* **2010**, *660*, 81–86. [CrossRef]

18. Revi, M.; Badeka, A.; Kontakos, S.; Kontominas, M.G. Effect of packaging material on enological parameters and volatile compounds of dry white wine. *Food Chem.* **2014**, *152*, 331–339. [CrossRef] [PubMed]
19. Clark, A.C.; Dias, D.A.; Smith, T.A.; Ghiggino, K.P.; Scollary, G.R. Iron (III) Tartrate as a potential precursor of light-induced oxidative degradation of white wine: Studies in a model wine system. *J. Agric. Food Chem.* **2011**, *59*, 3575–3581. [CrossRef]
20. D. P. R. 20.09. 1973. Assignment of the Denomination of Controlled Origin of "Malvasia delle Lipari" Wine Official Gazzette of the Republic of Italy n° 28, (30.01.1974). Available online: http://catalogoviti.politicheagricole.it/scheda_denom.php?t=dsc&q=2175 (accessed on 22 January 2018).
21. Muratore, G.; Nicolosi Asmundo, C.; Lanza, C.M.; Caggia, C.; Licciardello, F.; Restuccia, C. Influence of Saccharomyces uvarum on volatile acidity, aromatic and sensory profile of Malvasia delle Lipari wine. *Food Technol. Biotech.* **2007**, *45*, 101–106.
22. Belitz, H.B.; Grosch, W.; Schieberle, P. *Food Chemistry*, 3rd ed.; Springer: Berlin/Heidelberg, Germany, 2004; pp. 258–282.
23. Rada-Mendoza, M.; Olano, A.; Villamiel, M. Determination of hydroxymethylfurfural in commercial jams and in fruit-based infant foods. *Food Chem.* **2002**, *79*, 513–516. [CrossRef]
24. Rada-Mendoza, M.; Sanz, M.L.; Olano, A.; Villamiel, M. Study on nonenzymatic browning in cookies, crackers and breakfast cereals by maltulose and furosine determination. *Food Chem.* **2004**, *85*, 605–609. [CrossRef]
25. Arena, E.; Fallico, B.; Maccarone, E. Evaluation of antioxidant capacity of blood orange juices as influenced by constituents, concentration process and storage. *Food Chem.* **2001**, *74*, 423–427. [CrossRef]
26. Arena, E.; Muccilli, S.; Mazzaglia, A.; Giannone, V.; Brighina, S.; Rapisarda, P.; Fallico, B.; Allegra, M.; Spina, A. Development of Durum Wheat Breads Low in Sodium Using a Natural Low-Sodium Sea Salt. *Foods* **2020**, *9*, 752. [CrossRef] [PubMed]
27. Zappalà, M.; Fallico, B.; Arena, E.; Verzera, A. Methods for the determination of HMF in honey: A comparison. *Food Control* **2005**, *16*, 273–277. [CrossRef]
28. Spina, A.; Brighina, S.; Muccilli, S.; Mazzaglia, A.; Rapisarda, P.; Fallico, B.; Arena, E. Partial Replacement of NaCl in Bread from Durum Wheat (Triticum turgidum L. subsp. durum Desf.) with KCl and Yeast Extract: Evaluation of Quality Parameters During Long Storage. *Food Bioprocess Technol.* **2015**, *8*, 1089–1101. [CrossRef]
29. Spina, A.; Brighina, S.; Muccilli, S.; Mazzaglia, A.; Fabroni, S.; Fallico, B.; Rapisarda, P.; Arena, E. Wholegrain Durum Wheat Bread Fortified With Citrus Fibers: Evaluation of Quality Parameters During Long Storage. *Front. Nutr.* **2019**, *6*, 13. [CrossRef]
30. Pereira, V.; Albuquerque, F.M.; Ferreira, A.C.; Cacho, J.; Marques, J.C. Evolution of 5-hydroxymethylfurfural (HMF) and furfural (F) in fortified wines submitted to overheating conditions. *Food Res. Int.* **2011**, *44*, 71–76. [CrossRef]
31. Cutzach, I.; Chatonnet, P.; Henry, R.; Pons, M.; Dubourdieu, D. Etude de l'arome des vins doux naturels non muscatés. 2eme Partie: Dosage de certains composés volatils intervenant dans l'arome des vins doux naturels au cours de leur vieillissement. *J. Int. Scie de la Vigne et du Vin* **1998**, *34*, 211–221. (In French)
32. Lee, Y.C.; Shlyankevich, M.; Jeong, H.K.; Douglas, J.S.; Surh, Y.J. Bioactivation of 5-hydroxymethyl-2-furaldehyde to an electrophilic and mutagenic allylicsulfuric acid ester. *Biochem. Biophys. Res. Commun.* **1995**, *209*, 996–1002. [CrossRef]
33. Surh, Y.J.; Liem, A.; Miller, J.A.; Tannenbaum, S.R. 5-Sulfooxymethylfurfural as a possible ultimate mutagenic and carcinogenic metabolite of the Maillard reaction product, 5-hydroxymethylfurfural. *Carcinogenesis* **1994**, *15*, 2375–2377. [CrossRef] [PubMed]
34. Durling, L.J.K.; Busk, L.; Hellman, B.E. Evaluation of the DNA damaging effect of the heat-induced food toxicant 5-hydroxymethylfurfural (HMF) in various cell lines with different activities of sulfotransferases. *Food Chem. Toxicol.* **2009**, *47*, 880–884. [CrossRef]
35. Li, Y.-H.; Lu, X.Y. Investigation on the origin of 5-HMF in Shengmaiyin decoction by RP-HPLC method. *J. Zhejiang Univ. Sci. B* **2005**, *6*, 1015–1021. [CrossRef] [PubMed]
36. Rizzo, V.; Torri, L.; Licciardello, F.; Piergiovanni, L.; Muratore, G. Quality changes of extra virgin olive oil packaged in coloured PET bottles stored under different lighting conditions. *Packag. Technol. Sci.* **2014**, *27*, 437–448. [CrossRef]
37. OIV International Organization of Vine and Wine. Compendium of International Methods of Analysis of Wines and Musts (2 volUMES). Available online: https://www.oiv.int/en/technical-standards-and-documents/methods-of-analysis/compendium-of-international-methods-of-analysis-of-wines-and-musts-2-vol (accessed on 22 January 2018).
38. Di Stefano, R.; Cravero, M.C.; Gentilini, N. Metodo per lo studio dei polifenoli nei vini. *L'Enotecnico* **1989**, *Maggio*, 83–89.
39. Hopfer, H.; Ebeler, S.E.; Heymann, H. The combined effects of storage temperature and packaging type on the sensory and chemical properties of Chardonnay. *J. Agric. Food Chem.* **2012**, *60*, 10743–10754. [CrossRef] [PubMed]
40. Arapitsas, P.; Dalledonne, S.; Scholz, M.; Catapano, A.; Carlina, S.; Mattivi, F. White wine light-strike fault: A comparison between flint and green glass bottles under the typical supermarket conditions. *Food Packag. Shelf Life* **2020**, *24*, 100492. [CrossRef]
41. Díaz, I.; Castro, R.I.; Ubeda, C.; Loyola, R.; Laurie, V.F. Combined effects of sulfur dioxide, glutathione and light exposure on the conservation of bottled Sauvignon blanc. *Food Chem.* **2021**, *356*, 129689. [CrossRef] [PubMed]
42. Refsgaard, H.H.; Brockhoff, P.B.; Poll, L.; Olsen, C.E.; Rasmussen, M.; Skibsted, L.H. Light-induced sensory and chemical changes in aquavit. *LWT* **1995**, *28*, 425–435. [CrossRef]
43. Guerrini, L.; Pantani, O.; Politi, S.; Angeloni, G.; Masella, P.; Calamai, L.; Parenti, A. Does bottle color protect red wine from photo-oxidation? *Packag. Technol. Sci.* **2019**, *32*, 259–265. [CrossRef]
44. Muratore, G.; Licciardello, F.; Restuccia, C.; Puglisi, M.L.; Giudici, P. Role of Different Factors Affecting the Formation of 5-Hydroxymethyl-2-furancarboxaldehyde in Heated Grape Must. *J. Agric. Food Chem.* **2006**, *54*, 860–863. [CrossRef] [PubMed]
45. Cutzach, I.; Chatonnet, P.; Dubourdieu, D. Study of the formation mechanisms of some volatile compounds during the aging of sweet fortified wines. *J. Agric. Food Chem.* **1999**, *47*, 2837–2846. [CrossRef] [PubMed]

46. Fracassetti, D.; Di Canito, A.; Bodon, R.; Messina, N.; Vigentini, I.; Foschino, R.; Tirelli, A. Light-struck taste in white wine: Reaction mechanisms, preventive strategies and future perspectives to preserve wine quality. *Trends Food Sci. Technol.* **2021**, *112*, 547–558. [CrossRef]
47. Benucci, I. Impact of post-bottling storage conditions on colour and sensory profile of a rosé sparkling wine. *LWT* **2019**, *118*, 108732. [CrossRef]
48. Jung, R.; Kumar, K.; Patz, C.; Rauhut, D.; Tarasov, A.; Schüßler, C. Influence of transport temperature profiles on wine quality. *Food Packag. Shelf Life* **2021**, *29*, 100706. [CrossRef]
49. Lan, H.; Li, S.; Yang, J.; Li, J.; Yuan, C.; Guo, A. Effects of light exposure on chemical and sensory properties of storing Meili Rose wine in colored bottles. *Food Chem.* **2021**, *345*, 128854. [CrossRef] [PubMed]

MDPI
St. Alban-Anlage 66
4052 Basel
Switzerland
Tel. +41 61 683 77 34
Fax +41 61 302 89 18
www.mdpi.com

Foods Editorial Office
E-mail: foods@mdpi.com
www.mdpi.com/journal/foods

www.ingramcontent.com/pod-product-compliance
Lightning Source LLC
LaVergne TN
LVHW070905170726
843515LV00005B/707

* 9 7 8 3 0 3 6 5 2 6 3 7 9 *